PRESS BRAKE AND SHEAR HANDBOOK

Second Edition

A basic handbook on the design, selection
and use of press brakes and shears

Edited by
Harold R. Daniels
formerly Senior Associate Editor
Metalworking Economics

PUBLISHED BY CAHNERS BOOKS
Division of Cahners Publishing Company, Inc.
89 Franklin Street, Boston, Massachusetts 02110

Copyright © 1965, 1974 by Cahners Publishing Company, Inc.

International Standard Book Number: 0-8436-0815-3
Library of Congress Catalog Card Number: 74-13643

Printed in the United States of America

Preface

This book is a companion piece to the Mechanical Press Handbook published in 1960 by *Metalworking* magazine. Like its predecessor, it was written to fill the need for a basic manual on a particular subject. The subject, in this case, is of a dual nature. It was felt that press brakes and shears are in many ways complementary. Users of the former almost invariably also use the latter.

The *Press Brake and Shear Handbook* covers basic design considerations. It attempts to establish, for the man who must specify this kind of equipment, which of these considerations are important to him, and why. It also describes the many options that are available to the buyer of this kind of equipment and makes recommendations as to their use.

This second edition features a valuable discussion of safety procedures necessary to meet current OSHA requirements, and offers practical suggestions for implementing these federal standards. It contains a comprehensive safety manual for operators of mechanical press brakes, as well as many photographs and diagrams illustrating the wide applications of metal-forming equipment. The handbook also features a new chapter on recent developments in machinery controls and operating techniques.

Press-brake forming and shearing require a higher degree of operator skill and technique than does pressworking. For this reason, some guides to sound brake-forming and shearing practices have been included.

Before the job of compiling and editing the handbook was begun, virtually all press-brake and shear builders were asked to assist in the project. Without exception, their cooperation was generously given in the form of technical information, photographs and many other ways. The publishers wish to express their appreciation. The publishers also wish to acknowledge the special contributions of a number of individuals:

Mr. Robert L. Shelton, of Cincinnati, Incorporated, Mr. E. W. Steger of Dreis & Krump Mfg. Co., and Mr. R. H. Usinger of Pacific Industrial Mfg. Co. These men solved the selection problems in Chapter 2. In addition, Mr. Steger developed the material and supplied the drawings for the chapters on press-brake dies.

Mr. Howard S. Achler of Kaufman Tool and Engineering Corp. and Mr. C. H. Kuhns of Wapakoneta Machine Co. gave invaluable aid in the preparation of the chapters on urethane tooling and knives, respectively. Mr. H. E. Carson of Niagara Machine & Tool Works, Mr. Walter Johnson of Verson Allsteel Press Co. and Mr. A. W. Schultz of Cleveland Crane & Engineering Co. assisted in outlining the book and gave freely of their time in reviewing the material prior to publication. Their assistance is sincerely appreciated.

CONTENTS

MECHANICAL PRESS BRAKES

This chapter discusses the basic elements of the press brake and defines its potential and limitations. Some of the different approaches to press-brake construction are described and guidelines for proper press-brake selection are established in terms of the type of work to be done.

A press brake was originally conceived as a mechanism for making straight-line bends. Improvements in design and construction have greatly increased the range of work than can be done on it. The modern press brake is a versatile piece of equipment. Properly tooled, it can be used for blanking, piercing, punching, staking, drawing and many other applications.

Since the press brake can do so many jobs, the potential buyer of metal-forming equipment may ask, "Why shouldn't I buy a press brake instead of a press for my plant? After all, it costs a lot less than a press of similar capacity and bed length. And when I'm not using it for bending, I can stick in a progressive die and get some other work out."

There is no simple answer to such a question. The press brake is less expensive because it is primarily designed to bend on a long line and its construction is therefore simpler than that of a press. If it is used for additional functions, great care should be taken not to overload it.

There are many jobs that can only be done on a press. There are many jobs that can only be done on a press brake. There is also a wide area where a job can be done economically on either piece of equipment.

In this area differences of opinion exist even among respected authorities on metal forming. Perhaps the best advice would be, "Don't buy a press brake as a substitute for a press. Buy it as the useful piece of equipment that it is in its own right." Succeeding chapters will point out the

Lack of clutter contributes to safety of modern Niagara press brake.

wide range of press-brake operations and their relatively few limitations.

Just as in the case of presses, the press-brake buyer has a choice between mechanical or hydraulic equipment. This is a more difficult decision to make than is the case with presses. There is a wide overlapping area. Generally speaking, the hydraulic press brake is at its best on long-stroke, moderate-speed work. The mechanical press brake is faster, more accurate on some operations, and better suited to high-production operation.

The hydraulic press brake delivers its rated capacity over the entire stroke. The mechanical press brake is rated at mid-stroke but frame construction is usually such that 50% more than rated tonnage can be delivered at the bottom of the stroke. The hydraulic press brake has specific advantages that will be discussed in a later chapter of this series. Actual problems involving selection of hydraulic or mechanical brakes will be cited.

For a given bend on a given length of stock of a given thickness, it is theoretically possible to determine exactly what size and capacity press brake is needed. Even though the builders of press brakes offer an almost

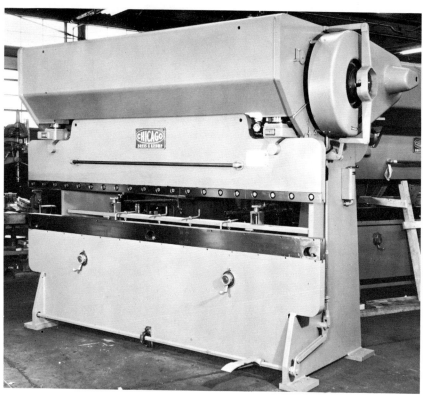

Massive construction, as on this Dreis & Krump brake, resists deflection.

unlimited range of equipment, press brakes are seldom specified on the basis of these three factors alone. Few press-brake buyers would wish to be limited to specifications established for one job. Further, very slight changes in any of the three factors mentioned can have a drastic effect on press-brake requirements.

Length and Capacity

The nominal length of a press brake is the distance between the inside faces of the two housings. In practice, most builders design their equipment so that the ram and bed extend beyond the housings. Usually a cutout or throat is provided so that the work can extend through the dies for a distance equal to the throat depth.

Work that passes inside the housings, of course, can be bent to any desired length. Thus, the specification of the length of a press brake obviously depends on the depth of the flange to be bent as well as the over-all width of the sheet.

Most press-brake builders furnish charts showing the tonnage needed to

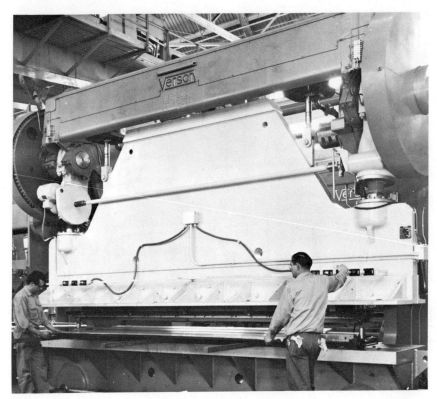

Dual control system protects operators of this large Verson brake. This one has ram and bolster extensions for punching application.

air-bend mild steel in a number of thicknesses over Vee dies in a range of openings. These charts are based on field reports and testing programs and they are remarkably consistent. Examination of the bending-pressure chart included with this chapter will bring out clearly how much the nature of the bend effects the required pressure.

For example, a standard bend will have an inside radius approximately equal to the material thickness when the Vee die opening is about eight times the material thickness. To reduce the inside radius by even a small amount, the Vee die opening must be reduced and the tonnage nearly doubled. If a larger radius is permitted, tonnage requirements drop off sharply.

This can be demonstrated, again referring to the bending tonnage chart, by three typical examples of common bends. Assume that the bends are made on $\frac{1}{4}$-in. × 10-ft mild steel with an average tensile strength of 55,000 to 65,000 psi.

Example 1. Standard inside radius from standard die opening. $\frac{1}{4}$ in.

Foot control switch on this Cincinnati press brake is guarded to prevent accidental accuation. Indicator lights are extra insurance.

over 2-in. die opening requires 15.4 T/ft. 15.4 T/ft × 10 ft = 154 tons. Inside radius is about $\frac{5}{16}$ in.

Example 2. Smaller inside radius from smaller die opening. $\frac{1}{4}$ in. over $1\frac{1}{2}$-in. die requires 22.7 T/ft. 22.7 T/ft × 10 ft = 227 tons. Inside radius is about $\frac{15}{64}$ in.

Example 3. Larger inside radius from larger die opening. $\frac{1}{4}$ in. over 3-in. die requires 9.0 T/ft. 9.0 T/ft × 10 ft = 90 tons. In this case, the inside radius is about $\frac{15}{32}$ in.

The tonnage required to make a bend is also a function of the tensile strength of the material being bent.

Specifications of a press brake of the proper capacity for a given range of work is thus dependent on the nature of the bends to be made and the materials to be formed. An ample reserve capacity should be allowed.

Other Factors

There are many other factors that should be considered in specifying a press brake. Thickness of the metal to be bent is one of these. A press brake rated for a given thickness of metal can bend much heavier metal if the allowable radius is larger or the length of the piece is reduced. Inversely, multiple small-radius bends in light-gage metal can sometimes demand more pressure than single bends in heavier gages.

To further complicate the specification problem, some press-brake builders express capacity in terms of tonnage alone. If this is done without taking flywheel energy requirements into consideration, the capacity rating can be misleading. Fortunately, most press-brake builders express capacity in terms of working pressures over Vee dies. This, of course, includes both tonnage and flywheel energy.

Design and Construction

A press brake is designed to deliver pressure over a long narrow area with a minimum of deflection. Because this is basic to all press brakes, most builders offer designs that are outwardly quite similar in appearance. This is not a reflection on the imagination and skill of the builders.

An analogy might be an axe. The shape of the axe is ideal to its purpose. It is impossible to conceive of any drastic variation in the shape of an axe. Since the function of the press brake is equally straight-forward, most press brakes will resemble each other and, in fact, are similar in design.

Builders do offer improvements and variations in basic design but it is in the area of drives, clutches, controls, open or enclosed gearing, optional equipment and auxiliaries that the great differences are found. These will be discussed in subsequent chapters.

The three major structural parts of a press brake are the frame, bed and ram. Almost without exception, American builders have settled on welded steel construction. Many builders, particularly in their smaller units, weld the bed to the frame. This makes for rigid construction provided that the entire frame is properly stress-relieved. Several major builders use interlocking construction with hand-scraped bearing shoes to attach the bed to the frame, particularly on heavier units. On some enormous press brakes, shipment by common carrier would otherwise be impossible. Other things being equal, the bed of a press brake should be deep and heavy to reduce deflection.

The ram should be of equally heavy construction. Ram and bed should be parallel under no-load conditions. This point will be discussed in detail in a section on crowning of filler blocks in a later chapter. If the press brake is to be equipped with a wide ram bolster and angle brackets to increase its range of use, heavy construction is essential. This point will be amplified in the chapter on dies. It is far more economical to over-specify than to underspecify when ordering any press brake.

Tonnage Requirements for Air Bends in Mild Steel

Metal Thickness		Width of Vee Die Opening in Inches																						
Gauge	Inches	1/4	5/16	3/8	7/16	1/2	5/8"	3/4	7/8	1	1 1/8	1 1/4	1 1/2	2	2 1/2	3	3 1/2	4	5	6	7	8	10	12
20	0.036	3.1	2.3	1.7	1.4	1.1																		
18	0.048	5.3	4.0	3.0	2.5	2.2	1.7																	
16	0.060	9.6	7.1	5.6	4.5	3.8	2.8	2.2	1.8	1.5														
14	0.075		11.9	9.2	7.6	6.3	4.7	3.5	3.0	2.5	2.1	1.8												
12	0.105				16.7	13.1	9.7	8.0	6.5	5.6	4.6	4.1	3.2											
11	0.120					19.2	14.2	11.1	9.0	7.5	6.3	5.5	4.4	2.9										
10	0.135						18.6	14.5	11.9	9.9	8.5	7.3	5.8	4.0										
3/16	0.188							27.4	23.1	19.3	16.4	14.3	11.2	7.5	5.7	4.4								
1/4	0.250									39.4	33.3	29.5	22.7	15.4	11.4	9.0	7.4	6.1						
5/16	0.313											50.4	39.8	27.0	19.7	15.3	12.7	10.5	7.7					
3/8	0.375												61.6	42.3	30.9	24.0	19.6	16.3	12.3	9.5				
7/16	0.438													61.7	45.8	35.4	28.6	24.4	17.3	14.8	11.2			
1/2	0.500													85.2	63.6	48.8	39.7	33.3	24.6	19.4	15.9	13.1		
5/8	0.625														110.0	86.2	70.0	58.3	43.1	33.3	27.4	23.3	16.9	
3/4	0.750															138.0	110.0	93.0	68.7	53.5	43.6	36.5	27.1	21.0
7/8	0.875																165.0	137.0	104.0	80.7	64.6	52.9	39.7	31.6
1	1.000																	197.0	143.0	113.0	91.2	76.2	56.3	44.2

Bed on big Steelweld brake is detachable. Jib crane is for heavy plate.

Most large press brakes are provided with throats in the side frames so that work wider than the housings can be bent. When the ram of the press brake descends, great stresses occur across the face of the throat. The forward edge of the frame is subjected to a strain in tension while the back of the frame is compressed. The throat should be provided with ample radii since sharp radii act as stress raisers. Most frame failures occur at the throat. One manufacturer uses a forging in this area. In specifying a press brake with deeper than standard throats in the side members, attention should be given to the type of construction used to minimize stress.

Guiding the Ram

The means by which the ram is guided is basic to the construction of the press. Major press-brake builders have taken different approaches to guiding the ram. The following examples are typical:

The Cleveland Crane & Engineering Co., in its Steelweld line, uses Vee-shaped ram slides to provide full bearing contract and prevent loose

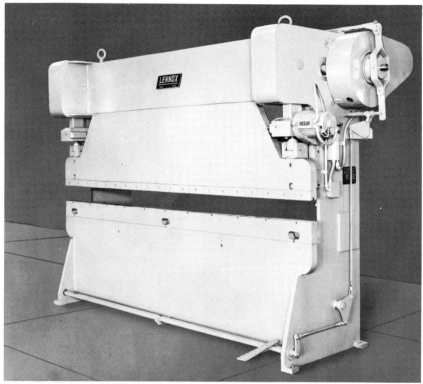

Capacity of this Lennox mechanical press brake is 10 ft, 14-ga mild steel.

gibbing. Front-to-back and side wear is self compensating. Shims are provided under the bearing caps on each side of the slide for wear compensation. Adjustment for slide wear can be made in a matter of less than half an hour.

Niagara Machine & Tool Works uses laminated non-metallic ways in all its press brakes, both mechanical and hydraulic. Exposed fiber ends act as wicks to retain lubricants. They can be quickly replaced. Rocker-type bronze end-bearings and ball-and-socket connections maintain alignment when the ram is tilted for tapered work.

Dreis & Krump Mfg. Co. also uses non-metallic liners on the rams of all its press brakes. Ball-and-socket connections are used in most models. The housing gibs are bolted to the frame.

To meet the alignment problem on tapered work, Cincinnati, Incorporated builds two large bronze shoes into the right-hand ram slide. Since the ram slide is free to rotate about the shoe, alignment is accurate regardless of the ram position.

Verson Allsteel Press Co., in its Major line, uses long square body gibs

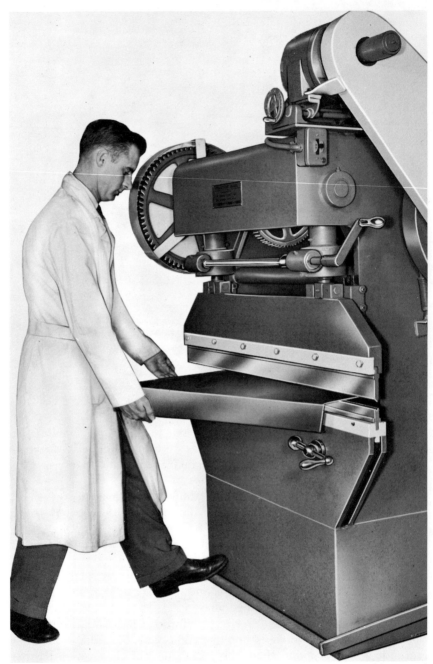

Connecticut (W. Whitney Stueck, Inc.) 24-ton press brake has variable speed drive; many other features usually found in larger equipment.

that confine the bronze ram gibs at all points on the stroke. Verson also uses ball-and-socket connections.

In the small press-brake area, W. Whitney Stueck uses double gibs, straight outside gibs and angle-type inside gibs on its Connecticut line of 24-ton press brakes.

The point to be established here is that, once the area of basic design is departed from, press-brake builders begin to incorporate their own engineering concepts into their product. This will be emphasized in the discussion of drives and controls that will appear, as will ram adjustment, in a later chapter.

Chapter 2

HYDRAULIC PRESS BRAKES

This chapter discusses the advantages and limitations of hydraulic press brakes. Some typical forming jobs are described and the recommendations of leading press-brake authorities as to type of equipment are given, along with their reasons for specifying hydraulic or mechanical equipment.

Chapter 1 of this book pointed out that the press-brake user could choose between two types of equipment: the mechanical press brake and the hydraulic press brake. It further stated that the distinction between the two types was by no means as clear cut as in the case of hydraulic versus mechanical presses. Recent developments in hydraulic-press-brake design have eliminated some of the traditional advantages of the mechanical press brake. Inversely, new control systems for mechanical press brakes have cut into the domain of hydraulic press brakes.

In the final analysis, the choice between the two types should be based on the nature of the work to be performed. It should be recognized that hydraulic press brakes still retain some inherent advantages as do their mechanical counterparts.

Advantages—Hydraulic Equipment

Perhaps the biggest single advantage is that a hydraulic press brake, under ordinary circumstances, cannot be overloaded to the point where damage to the die or to the press brake itself will result. The tonnage of a hydraulic press brake is a function of the size of the cylinders and pump and of the capacity of the circuitry. This is a known and fixed quantity. The construction of the press brake itself is matched to it.

Since the fixed tonnage cannot be exceeded, a hydraulic press brake can be bottomed at full tonnage repeatedly without hazard. Further, re-

Enormous Verson press brake symbolizes raw power of hydraulic equipment. This one develops 4-million lb pressure, yet it has a highly sensitive control system with four pressing speeds. The bed on this machine measures 26 ft × 2½ ft.

gardless of the size or thickness of the work, the ram will stop when it reaches the designated tonnage. It can be backed away from any point on the stroke.

Because of these characteristics, relatively unskilled operators can be safely assigned to a hydraulic press brake with only a remote possibility that they can damage equipment through lack of skill. The operator retains control of the ram at all times. He can inch the press brake in infinitely small increments if desirable. This is a definite asset in setting up new dies.

Simple stroke adjustment is another advantage of the hydraulic press brake. It is usually accomplished by mounting limit switches at the desired top and bottom limits of ram travel. Almost all models marketed by U. S. builders are equipped with a micrometer adjustment for setting the exact position of the ram at bottom-of-stroke. The ram can be positioned to within a thousandth in many cases. By this means, a repeat job going into the press brake can be set up to produce identical parts in a matter of a few moments.

A hydraulic press brake delivers full rated power throughout the stroke. With certain dies, particularly those with a high bottom section as is necessary on certain bends, this is an important advantage. A mechanical press brake, since it uses energy stored in a flywheel, cannot duplicate this characteristic. Hydraulic press brakes, since they are not limited by crankshaft design factors, are available with extremely long strokes. For economic reasons, of course, the stroke should be kept as short as possible for a given job to eliminate unnecessary ram travel.

Some Limitations

Other things being equal, a hydraulic press is not as fast as a mechanical press brake of similar tonnage and size. Design improvements in modern hydraulic press brakes, however, have greatly reduced this disadvantage. Most hydraulic press brakes have at least two speeds and sometimes three.

The first speed is that used when the press brake is operating at or near full tonnage capacity. On a typical modern press brake, a faster speed— about double the basic speed—is used to approach the work and on the return stroke of the ram. This faster speed can be used throughout the stroke on work where very light tonnage is required. Obviously this increases productive capacity.

On the model under discussion, a third speed is provided. About four times faster than the basic speed, it cuts down approach and return time. The operator sets up the cycle by adjusting a cam interlocked with the control limit switches.

Excessive maintenance and leakage are often cited as objections to hydraulic press brakes and shears. It is true that earlier models did leak

Smaller Pacific hydraulic brakes demonstrate vastly improved versatility of hydraulic equipment. These are a matched pair. First machine punches holes in copper connectors. Second brake is used for forming. Plating bath separates the two jobs.

and frequent replacement of packing and seals was necessary. In the manufacture of modern equipment, greater attention is paid to cylinder finishing and valve design. Chevron-type seals and better O-ring materials have reduced the maintenance problem so that it is no longer a valid selection factor.

Various manufacturers of hydraulic press brakes offer a range of speeds and a great many options for their precise control. Satisfactory production speeds on long runs are easily obtained by fast approach and return speeds combined with slow work speeds.

Tonnage Control

A useful feature of modern hydraulic press brakes is a preset tonnage control. This function is usually infinitely variable within the capacity of the press brake. It makes it possible to use inexpensive dies for light bending work.

Epoxy and polyurethane dies have given satisfactory service on such applications. In bending polished or prefinished stock, the stroke can be adjusted to be as light as possible and still make the necessary bends. This reduces the possibility of scuffing or otherwise marking the stock.

When the nature of the work is such that light tonnage is required, it is often possible to operate the press brake at a higher speed rate.

In preparing this chapter, some outstanding authorities in the field of press-brake design were asked to answer some hypothetical problems concerning press-brake selection. It will be noted that there is considerable difference of opinion. This was to be expected and actually adds to the value of the answers. In justifying their opinions, the correspondents have provided a wealth of guidance in the area of press brake selection.

Mr. Brown's Problem

Mr. Brown manufacturers steel shelving from prefinished 18-ga mild steel. His plant works two shifts per day. Shelves are made in many lengths to 72 in. and in three widths. There is a standard radius on one edge and a 180-deg curve on the other. The shelves are later assembled with bolts. Three basic hole patterns are used. Mr. Brown wants to punch these out on a press brake. He now uses press brakes but needs additional output of 8,000 additional pieces per day. He is considering roll forming. His existing equipment cannot keep up with his needs. What type of additional equipment would you recommend that he buy? Hydraulic or mechanical?

Solution...E. W. Steger, Dreis & Krump Mfg. Co.

"This work could readily be done on a mechanical straight-side press equipped with feed and transfer mechanisms. The material could be fed into the press from a stock coil and into the pierce and shear position.

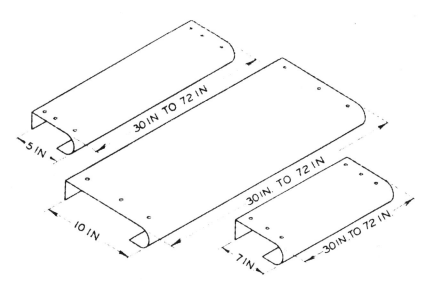

Mr. Brown's Problem. Shelving is made in three widths in a number of different lengths. Further, hole spacing is not the same for all widths.

The transfer mechanism would move the sheared and pierced section into the forming position. It could be set up to move the formed piece out onto a conveyor. All tooling would be adjustable for width and length so that only one set of tooling would be required. Templets could be used to reset the piercing and shearing units. The forming units could be relocated by positioning into prelocated slots and bolting down solidly. Production would be from 11 to 15 parts per min, depending on the length of the part being handled. Roll forming equipment would prove to be quite costly for this job because of high initial cost, the necessity for various roll widths and the time element involved in changing rolls. This does not appear to be an ideal job for a press brake."

Solution. . .R. H. Usinger, Pacific Industrial Mfg. Co.

"Mr. Brown's work would be done on a single machine, using progressive dies. I would suggest an 18-ft, 100-ton hydraulic press brake with a fixed 12-in.-wide lower platen and a high-speed power unit. The ram should be machined for the addition of platens and I would suggest a width of 12 ft between housings. This would mean that the press would have two 24-in. horns to give the required length to handle three 72-in. sections. The brake would be set up with three operators with the first operator punching nine holes in a single stroke. He would probably use an equivalent of a Whistler die set to allow fast changeover for different hole spacings. The second operator would form the 180-deg radius and the third would form the 90-deg bend.

"Length of stroke should be 4 in. with 3 in. of rapid advance and 1 in. of pressing. The operation could be speeded up by reducing stroke length but this would probably increase handling time disproportionately. Since the clutch of a hydraulic press cannot be slipped, whip-up would be a problem. Press speed would have to be slowed down so that advance speed is 250 imp with a pressing speed of 35 ipm. Actual cycle time would then be a little less than 5 sec. Allowing 13 sec for handling, actual production rate would be about 200 per hr. At this rate, working two shifts, about 3,000 parts per day could be made. This might be boosted to 4,000 per day so that Mr. Brown would only need two presses. In the long run, he would probably be wiser to use three press brakes. Three presses could be purchased for about $45,000. Using Whistler dies, different parts could be run without die changes.

An alternative method would be to use a group of light mechanical press brakes. The advantages of using progressive tooling would be lost if this were done. Further, much more floor space would be required. Actually, an 18-ft mechanical press brake could be used for Mr. Brown's job. The mechanical press brake would cost more money but it would provide the advantage of being able to slip the clutch. I doubt if roll forming could be used profitably on Mr. Brown's job because of high initial cost and lack of flexibility."

Solution . . . Robert L. Shelton, Cincinnati, Incorporated

"The shelving could most economically be made in three successive hits —punch, form 180-deg radius, and wipe down 90-deg bend. Tooling could be set side-by-side on the bed of a press brake having an over-all die surface of 20 ft. Brake capacity of about 200 tons would be adequate. With the tooling set side-by-side, a complete shelf would be made with each stroke of the press brake. Three operators would be needed, passing the work from one station to the next. With such a setup, a speed of about ten pieces per min would be maintained. On this basis, the press would produce 4,800 shelves per 8-hr day. Since 8,000 shelves daily are required, Mr. Brown would need two machines in order to get required output in one shift. Because of the speed requirements, a hydraulic press brake would be impractical.

Mr. Smith's Problem

Mr. Smith owns a job shop. He is reluctant to spend money for equipment and usually buys used machines. He works for the most part with $\frac{1}{4}$-in. to $\frac{3}{4}$-in. steel on a variety of short-run jobs. His "bread and butter" is a contract to make truck body parts from $\frac{9}{16}$-in. mild steel plate. These have simple 90-deg bends

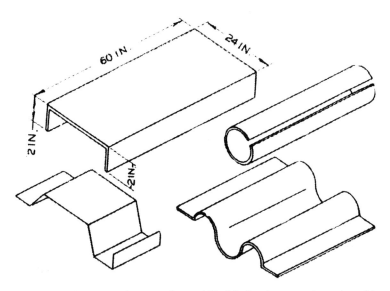

Mr. Smith's Problem. He has mostly unskilled help; has one bread-and-butter
job. (top.) Rest are typical short-run jobs from 36 in. to 10 ft in length.

and he makes 400 parts a day, all the same pattern. He pays low wages, has
relatively unskilled help. He wants to run the bread-and-butter job in the morn-
ings and run short-run jobs in the afternoons. These run under 150 pieces and
range from simple flanges to tubing. What equipment do you recommend to
Mr. Smith?

Solution. . .E. W. Steger, Dreis & Krump Mfg. Co.

"Mr. Smith should seriously consider a hydraulic press brake because
of the range of his tonnage requirements. The press should be equipped
with fast advance, normal pressing, slow pressing, fast return with slow
start, and a turret depth stop. The slow pressing will help prevent whip-up
during forming. The slow start of return will let the plate lower gradually
before the ram leaves the work. The turret depth stop will permit using the
same tooling on a number of jobs and eliminate much die changing.

"Tube forming could be done using an angle-forming punch in an air-
forming die. Rib forming could be done in three operations also using
angle form tooling. Three operations are recommended because of tonnage requirements in the heavier plate sizes. Corrugation forming
should also be an air-bending operation to cut down tonnage needs. To
form the 5-ft $\times$ $\frac{9}{16}$-in. plates will require 270 tons, air bending, one bend
at a time.

"I would recommend a hydraulic press of between 400- and 500-ton capacity. This is based on a review of the jobs that are being done in Mr. Smith's shop. The corrugating of $\frac{1}{4}$-in.-thick × 5-ft mild steel plate could then be conveniently handled with the probability of welding two formed pieces together to form a 10-ft section. If the sections were to be formed to the desired length of 10 ft, approximately 950 tons would be needed, assuming that the corrugations would have approximately 4-in. centers. Welding two short pieces together will require less initial investment in equipment."

Solution. . .R. H. Usinger, Pacific Industrial Mfg. Co.

"Mr. Smith could use either a mechanical or a hydraulic press brake. I would recommend a hydraulic. Mr. Smith buys used machinery and is obviously price conscious. The price of a 500-ton hydraulic press is less than that of a mechanical press brake. Further, he uses unskilled help. It is impossible to jam or damage a hydraulic machine. Unskilled personnel would not endanger the dies or the brake itself. Mr. Smith needs inexpensive dies because he has short-run jobs. By using a hydraulic press brake with tonnage control, less expensive dies can be used since they do not have to absorb the full rated tonnage. Further, the hydraulic press brake has a very wide adjustment of shut height and it is usually easier to adapt any type of die to a hydraulic than to a mechanical press brake.

"Smith's production runs are low. Setup time is probably more important than operating time. A hydraulic press brake is easier to set up than a mechanical press brake. Finally, Mr. Smith bends pipe occasionally. To close the pipe requires a rather long stroke, characteristically an advantage of hydraulic equipment that is also useful when forming deep box-type sections. I would recommend to Mr. Smith that he buy a 500-ton hydraulic press brake with 12 ft between housings."

Solution. . .Robert L. Shelton, Cincinnati, Incorporated

"The $\frac{3}{4}$-in.-thick × 10-ft-long mild steel used on some of Mr. Smith's short-run items would require a 34 Series Cincinnati mechanical press brake rated at 750 tons at bottom of stroke or a 500-ton Cincinnati hydraulic press brake rated at 500-tons throughout its stroke. If the $\frac{3}{4}$-in. plate is bent over an 8-in.-wide die opening, the 500-ton hydraulic brake would be adequate and considerably less expensive than the mechanical press brake. Mr. Smith wants to make 400 pieces, each with two bends in 4 hr. The $\frac{9}{16}$-in. material in his bread-and-butter job would probably be done over a $4\frac{1}{2}$-in. die opening, which would call for about a 4-in. stroke on the hydraulic machine. At this combination, the 500-ton hydraulic press brake would be capable of 4.2 spm—fast enough for the job.

"If Mr. Smith's tube job calls for high accuracy requiring coining at the

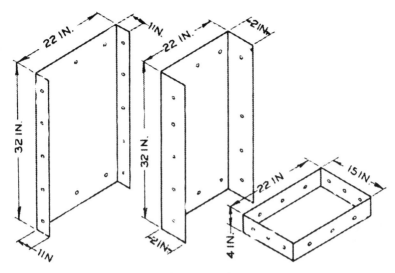

Mr. Jones' Problem. He makes furnace cabinets; needs additional capacity. Spacing of holes is critical. They're punched 9/32 in. for ¼ -in. bolts.

bottom of the stroke, a 1500-ton press brake would be required even if the tube were only $\frac{1}{4}$ in. thick × 10 ft long. If a very accurate tube of $\frac{3}{4}$-in. material was required, the tonnage requirements would be far higher. If accuracy is not stringent, the tube can be formed by successive bumps or hits with simple tooling."

Mr. Jones' Problem

Mr. Jones' plant makes furnaces. There are six basic parts. There are two sides, front and back, and top and bottom. Basic material is mild steel, 16-ga. The two side pieces have a 2-in. flange on each side. The front and back have 1-in. flanges on each side. All radii are standard. Sheets are preblanked on a straightside press and holes are gang-punched. Production is 100 furnaces per day. Mr. Jones wants to form the needed 600 parts during the morning shift and convert to short-run operations in the afternoons. The economics of press-brake forming look good to him but he needs convincing. What equipment do you recommend?

Solution. . .E. W. Steger, Dreis & Drump Mfg. Co.

"A wide mechanical press brake with a wide bed and ram or a mechanical straight-side press should be considered for Mr. Jones' furnace cabinet job. Three sets of channel dies would be set into the machine: one set for the front and rear sections, one set for the side sections, and one set for the top and bottom sections. Three sections would be formed with each stroke of the press, thus all the sections for one cabinet would be formed with two press strokes. To make 100 cabinets would require 200 press strokes, which could easily be done in 3 to 4 hr. This would allow from 54 to 72 seconds for loading, forming, and unloading the press."

Solution. . .R. H. Usinger, Pacific Industrial Mfg. Co.

"Mr. Jones' job could be done on either a mechanical or hydraulic press brake. I would suggest that progressive tooling be used. The press brake should be equipped with wide platens. A single die should be used to blank out the work and to punch the holes. The work could be done on a straight-side press but a press brake would be less expensive. I would suggest that steel rule dies be considered for the blanking operation and a Whistler die set for the hole punching. If the volume is sufficient, these two operations could be combined.

"I think that Mr. Jones should buy a 12-ft press brake and set it up for blanking and punching at one end of the machine and use a conventional Vee die for bending at the other end. He might also put in a third set of box-type dies so that the box could be closed in the third station. In this way, the box could be formed from beginning to end on a single machine with one operator who could move from station to station. Because of the low tonnage, I would recommend a 100-ton press brake with 30-in. wide platens. The lower platen should be fixed and permanently built into the machine while the upper platen should be removable. I would suggest an upper platen 36 in. long. We would then have a conventional Vee die set 36 in. long for the Vee bending and a third box-type die 16 in. long for closing the boxes.

"The economic problem is whether to buy a mechanical or a hydraulic press brake. I would recommend the hydraulic press brake because of its lower initial cost although from the production standpoint the mechanical press brake would be faster. In either case, mechanical or hydraulic, the tooling I have suggested would deliver good performance at low cost."

Solution. . .Robert L. Shelton, Cincinnati, Incorporated

"Mr. Jones' problem involves relatively simple tooling. One set of furnace parts would require 16 bends. One hundred furnaces would require 1,600 press strokes in 4 or $3\frac{1}{2}$ hrs, allowing a half hour for die change and setup to do the bread-and-butter jobs during the afternoon. To do 1,600 hits in $3\frac{1}{2}$ hrs would require eight hits per min. The tonnage requirements are very low and could easily be met by a 4 Series Cincinnati press brake rated at 65 tons on the front quarter and 100 tons at bottom of stroke. Because of the low tonnage requirements there is no need to go to a hydraulic machine and the high speed (40 spm) of the Cincinnati mechanical press will be a useful asset on the long-run, bread-and-butter afternoon work.

"This does not mean that low tonnage requirements always indicate a mechanical press brake. We recently delivered a hydraulic press brake tooled for making deep ribs in transformer tanks. The material was 20- to 16-ga mild steel and the ribs ranged in depth from 4 to $9\frac{5}{8}$ in. They ranged in pitch from $1\frac{1}{2}$ to $2\frac{3}{4}$ in. and were about 36 in. long. The job required a 400-ton Cincinnati hydraulic press brake with a 30-in. stroke— impossible on any mechanical press brake. Here was an application that clearly indicated a hydraulic press brake even though the tonnage requirements were low."

Chapter 3

MECHANICAL PRESS-BRAKE
DRIVES AND CONTROLS

This chapter discusses some of the many approaches press-brake builders have taken in developing drives and controls for their equipment. Particular attention is paid to new control methods that can often increase the productivity of a press brake and improve the quality of its output.

At about the 75-ton level, rather elaborate drive and control refinements are needed for mechanical press brakes. For that reason, this chapter will be devoted to press brakes of about that tonnage. A later chapter will be devoted to smaller press brakes and their operating characteristics.

Motors and Drives

The train of power for a mechanical press brake begins, of course, with the motor. In general, a press brake rated at 75 tons near the bottom of the stroke requires a motor rated at about 5 hp. Assuming that a press brake rated at 300 tons requires 15-hp motor and a press brake rated at 600 tons near the bottom of stroke a 30-hp motor, motor horsepower for the intervening sizes can readily be interpolated. For special purposes, either lighter or heavier motors can be furnished. Electrical manufacturers have developed families of motors specifically designed for use on presses and press brakes.

Most press brakes in the sizes under discussion have main drive gears at each end of the drive shaft. The purpose of this design is to assure that the ram descends parallel to the gibs regardless of the loading condition.

Phantom view of Dreis & Krump brake showing herringbone gear drive, one-piece eccentric and gear, strap, and plunger. All are enclosed in an oil bath.

This is essential in many types of press-brake work where off-center loading conditions normally exist.

Double-reduction gearing is usually furnished on press brakes from 75 tons upward. Typical is the construction used in Cleveland Crane & Engineering Co.'s Steelweld line. These brakes use two-stage reduction gearing. The first reduction runs in an oil bath in a sealed case. The bull gears are driven by pinions mounted on a power shaft extending across the width of the press brake. The bull gears are keyed to the eccentrics, which in turn drive the ram through the pitmans.

Dreis & Krump Chicago press brakes are made with oversized crankshafts. Pinion teeth are cut in the shafts and the eccentric gears rotate on fixed pins. This reduces the number of parts in the drive to a minimum and lowers potential maintenance costs.

Another drive variation is that offered by Verson Allsteel Press Co. in its Major line. On these brakes the flywheel is mounted on a quill rather than on the crankshaft. This eliminates a possible hazard inherent in driveshaft-supported flywheels.

Regardless of variation in drives, use of steel gearing is universal. This is essential because of shock loadings to which a mechanical press brake may be subjected. Some builders furnish forged steel pinions. Also avail-

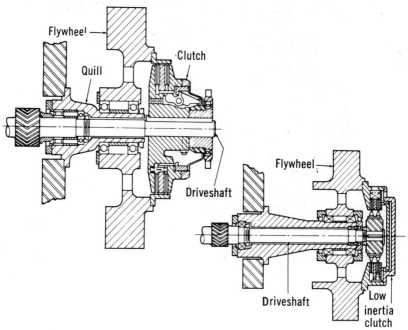

Quill-mounted flywheel and friction clutch are used in a number of Verson press brakes through Series B-17. Low inertia clutch is for larger models.

able are herringbone gears and precision hobbed spur gears. Because of the attention given to this detail by builders, gear failure on a modern press brake is extremely rare.

Precision gears contribute to the smooth, quiet operation of the modern press brake. Also a contributing factor is the use of antifriction bearings. With the exception of main bearings and slides, use of such bearings is almost universal.

Most modern mechanical press brakes are furnished with two-speed drives, either as an option or as standard equipment—usually as standard on the larger sizes. This commonly takes the form of a simple, lever-controlled gear box. The low-speed drive is used when bending wide sheets to reduce the amount of whip-up. A variation on the two-speed drive, offered by one builder, is the use of a variable speed drive that provides an infinite range of speeds between 20 and 50 spm.

Ram Adjustment

Motorized and tilting ram adjustment should be specified in buying new equipment in the medium to heavy range. It is an invaluable time-saver in fade-out work and in setting up repetitive jobs. It can also be used to overcome defects in tooling and dies.

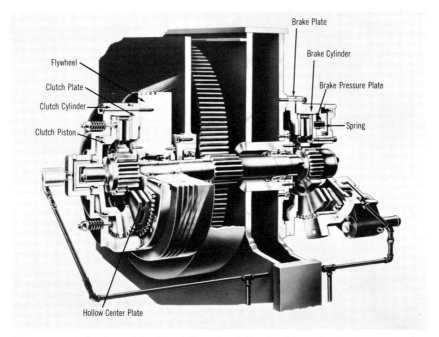

Air-operated clutch and brake of Cleveland Steelweld mechanical press brake. It's standard on models AN and larger, optional on other brake sizes.

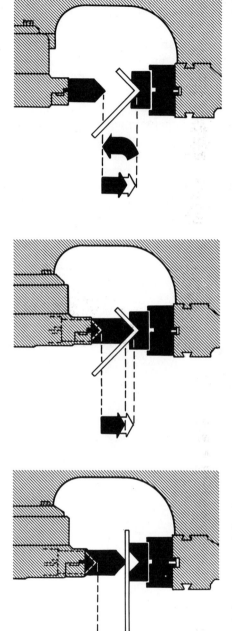

When the bend is completed, the ram returns to top dead center at high speed automatically. Since the clutch does not have to be slipped, clutch wear can be greatly reduced.

Just before it contacts the work, the ram slows. It continues at low speed to bottom dead center. Whip-up is greatly reduced at no greater sacrifice in lost operating time.

Cincinnati, Incorporated brake with automatic ram cycle boosts production of formed parts. Ram runs down at high speed until it is just above the work. Black arrow shows high-speed travel.

This feature is standard on most modern press brakes. It is usually accomplished with an independent motor connected, through a worm-gear drive, to a pair of adjusting screws inside the pitmans. The tilting is accomplished by several means, including split couplings and clutches that make it possible to disengage the drive on one side of the ram independently.

A variation is the use of bronze nuts located inside the pitmans. With this arrangement, the screws do not rotate. The worm-driven nuts move the ram up and down. This is a somewhat more expensive arrangement but it completely eliminates cramping while providing independent movement of either side of the ram.

Regardless of the type of ram adjustment, indicators should be provided at each end of the ram. They should be calibrated, preferably in thousandths, to show the exact position of each side of the ram and thus make it possible to set up a repeat job with a minimum of down-time and lost motion.

Remote control of ram adjustment is an inexpensive option that is a valuable aid to the toolsetter. It speeds die setting and makes it unnecessary for the operator to control ram adjustment during the process. He need not even be present when dies are changed. In addition to decreasing manpower waste, this has a positive safety value.

Clutches and Brakes

Most press-brake builders furnish mechanical disk-type clutches. Because slipping of the clutch is necessary on many press-brake operations, builders have developed and refined them to a high degree of reliability.

Combination friction clutches and brakes with interchangeable parts are widely used. Regardless of make, the clutch should provide the operator with continuous and positive control of the ram. It should be easily accessible for maintenance and service. Although they are remarkably reliable, press-brake clutches are subjected to the hardest kind of usage.

Typical of modern clutch design is that furnished by Niagara Machine & Tool Works on its Series N press brakes. The Niagara clutch is a development of the heavy-duty friction clutch used on Niagara power presses. It is a low-inertia type of simple construction, designed to fail safe. Most of the weight of the unit rotates with the flywheel. The driving plate alone starts and stops in each cycle. By means of a pressure regulator, the torque output of the clutch can be controlled or adjusted precisely. In this sense, it serves to protect the brake and the dies against overloading.

Regardless of the type of brake used—disk or band—it should be of sufficient capacity to stop the ram instantaneously when the clutch is disengaged. It should be able to do this even under the most rugged operating conditions with frequent stops and starts.

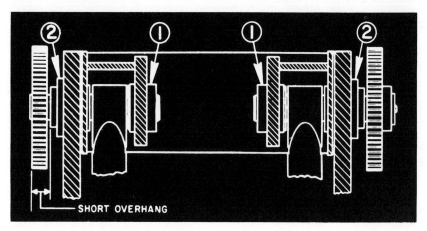

Eccentric shaft on Lodge & Shipley brake is one piece. Both shafts on this machine are supported on two full bearings to minimize friction.

Air Friction Clutches

The air friction clutch is actuated by expansion of a tube or diaphragm to force friction shoes against a hub or drum. There are many advantages to this type of clutch. It is exceptionally reliable. It has a cushioning effect that has shown some remarkable results in extending die life in press operations. It is easy to service and lends itself well to advanced control systems.

Until recently, the use of air friction clutches on press brakes was usually confined to a limited group of punching operations. There was a good reason for this. In bending metal on a brake, the operator usually slips the clutch just as the punch touches the work. If the ram is allowed to hit the work at full speed, dangerous whip-up takes place. The air friction clutch does not lend itself to this kind of work.

Exploded view of Wichita low-inertia air tube disk clutch. This type of clutch is used on a number of press brakes and mechanical presses.

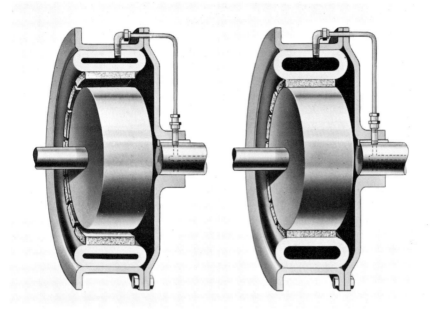

Drawing of widely-used Fawick Airflex clutch shows how tube expands to force friction surfaces against drum. Action is fast, smooth, and even.

Air friction clutches have now been successfully adapted to press-brake work by most press-brake makers working in cooperation with clutch builders. (Most air friction clutches, incidentally, are made by specialists. Fawick, Wichita and Industrial Clutch Corp. are major producers.) The Dreis & Krump Select-a-Speed is an excellent example. One of the first successful designs, this control uses a planetary-gear, 8 to 1 speed reducer enclosed within the flywheel to effect changes in ram strokes per minute through two electrically-controlled, air-operated clutches. The operator can select high-speed operation for the full cycle of the ram. He can select fast advance, low-speed bending, and high-speed return of the ram. A cycle control feature makes two types of operation available for each of the three speeds.

An interesting test, made with the Dreis & Krump equipment, indicates the potential of press brakes equipped with air friction clutches and advanced controls. In this case, a 45-ton standard press brake, operated by a skilled man, was used to form a metal pan measuring 24 x 27 x 1 in. deep from 16-ga steel. The tooling was a standard 90-deg Vee-die and the corner notches were pre-cut.

An inexperienced operator was given the same job, using a 45-ton press brake equipped with the Select-a-Speed feature. His brake was set so that

Replacement of worn disks on modern clutch is simple and quick.   First, remove bolts and insert guide pins.

Cylinder assembly can now be slid back on drift pins. Remaining cylinder bolts have already been removed.

Linings are in half sections on both sides of steel center plate of clutch. They slip in and out of place easily.

Clutch is now open with one lining disk removed and the other already in place. Unit is now ready for assembly.

Relining of brake is simple. Permanent slide pins are provided. Next step here will be replacement of linings.

Job is done. Clutch and brake shown are on Steelweld press brake. Modern clutches are easily maintained.

the punch came down to the bend line at 40 spm, automatically went into 5 spm to form the bend, and returned at 40 spm, automatically stopping at top of stroke.

The skilled operator, figuring loading time, forming all four bends, one at a time, and unloading, required 40 seconds to make one pan. The unskilled operator did the job in 27 sec. The 32.5% improvement in productive speed stemmed primarily from the fact that the operator did not have to slip the clutch on the Select-a-Speed unit, nor did he have to devote his attention to stopping the ram at or near top of stroke.

As mentioned, most major press-brake builders have adapted the air friction clutch to some form of multiple speed operation, usually by using dual clutches together with a speed reduction means incorporated into the flywheel. Production increases on the order of 30% and higher have been commonly attained. Another advantage is a high degree of uniformity in the bent product.

Air Electric Controls

The adoption of air electric control has greatly improved the flexibility, productivity and safety in operation of modern press brakes. Air electric control in a variety of forms is offered by all builders with and without air friction clutches. Usually it is an option with smaller machines and standard on larger brakes.

Basically, the air electric control replaces the mechanical foot treadle in operating the press brake. It makes it possible to locate the control station wherever desired. it can be set up with multiple stations in such a manner that several men can work safely on a long brake. The ram will not descend until the circuit is completed by each of the workers. It should be pointed out that devices permitting safe operation of purely mechanically controlled brakes are available. These, of course, do not provide the flexibility of the air electric type.

Almost any existing press brake can be equipped with air electric control. This is a growing trend in the industry for the reasons already stated and for the additional reason that such controls sharply reduce operator fatigue. They also make possible a degree of automation.

In installing air electric control, it is highly desirable to retain the option of operating the brake manually. There will always be heavy jobs where slipping of the clutch is required.

For long runs, semiautomatic and automatic operation of some press brakes is possible. Two approaches to automation will be discussed here. The first is that offered by Lodge & Shipley Co. This company has developed a unit disk clutch and brake that is self adjusting. The control includes a selector switch with three positions: jog, off and long. The jog position lets the operator start, stop or inch the ram at any point in the stroke. Holding the foot switch down provides continuous operation.

The long position allows him to jog the ram through 170 deg of its down stroke to a point 10 deg before bottom. From this point, the brake will automatically continue through the balance of the stroke, returning the ram to the top position. A safety device prevents double cycling. The off position bypasses the footswitch, making it impossible to trip the brake accidentally with the foot switch.

Cincinnati, Incorporated also has developed a control system that provides both fast and slow ram speeds during a single stroke. The ram starts from top of stroke at high speed and automatically shifts into low speed for the working portion of the stroke. At the bottom of the stroke, the ram again shifts to high speed for the return stroke. This again eliminates whip-up and back bending when forming long and flexible parts at high speeds without slipping the clutch.

The brake can be operated in six different ways by means of a control switch. There are short-dwell, medium-dwell and long-dwell positions. These positions regulate the length of the stroke during which the ram remains in slow speed. When the control is set at any of the dwell positions, all strokes are identical.

A high position on the switch allows the operator to jog the ram to bottom dead center in high speed. It then automatically returns to high speed for the return stoke. Low jog and high jog positions allow the ram to be jogged through the entire stroke in low and high speeds respectively. Major advantages of this system are reduction in operator fatigue by elimination of clutch slipping, a reduction in operator skill required, and sharply increased production. Up to 60% gains have been recorded.

There are a number of additional available refinements to the drive and control systems of modern mechanical press brakes. These include reversing ram drives, overload devices, air cushions and counter-balances, tonnage indicators and others.

Most major press-brake builders offer them as standard or optional equipment. The range of variations make it impractical to discuss them here, particularly since they can be considered as auxiliary, rather than basic equipment. However, they should be considered in specifying a press brake. Singly or in combination, they can increase the productivity of this versatile equipment.

Chapter 4

HYDRAULIC CIRCUITS AND CONTROLS

In selecting a hydraulic press brake, the buyer does not have as many options as he does when he selects a mechanical press brake. Nevertheless, there are some factors that he should consider in specifying this kind of equipment. This chapter discusses the more important options in detail.

The hydraulic press brake differs from the mechanical press brake in that fluid is used as a medium to transmit power. In a sense, the hydraulic cylinder performs exactly the same function as the crankshaft of the mechanical press brake.

Some basic knowledge of hydraulics is needed in order to evaluate properly the advantages and disadvantages of hydraulic vs mechanical transmission of power.

Modern hydraulic systems are all designed around a concept originated some three centuries ago by the French mathematician Blaise Pascal. Pascal's Law is the name given to this concept. It states that "pressure exerted on an enclosed liquid is transmitted undiminished in every direction."

Pascal developed his concept by experiments conducted with crude glass bottles and water. Now, 300 years later, it is being applied to enormously powerful machines. A hydraulic press brake built recently by Verson measures 44 ft between its uprights and weighs more than 1-million lb.

All hydraulic systems have four points in common. They must provide a reservoir for the medium, a pump, one or more valves and a cylinder. This is true whether referring to a fork lift truck, a hydraulic press brake or any other mechanism using hydraulic power.

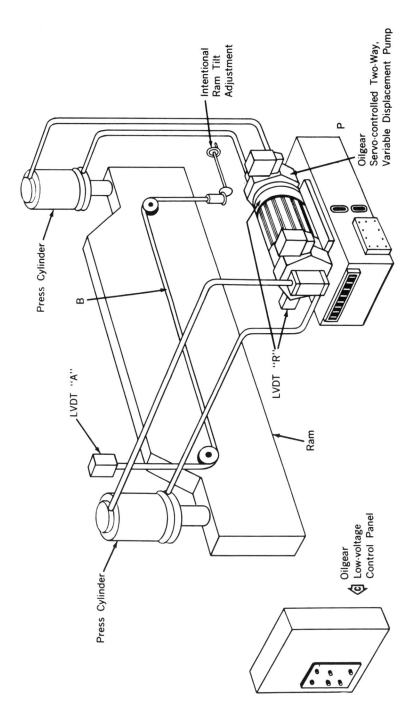

Intentional Ram Tilt Adjustment

Press Cylinder

B

LVDT "A"

LVDT "R"

Ram

Press Cylinder

Oilgear
Low-voltage
Control Panel

P

Oilgear
Servo-controlled Two-Way,
Variable Displacement Pump

Basic circuit made by Oilgear Co. for Dreis & Krump Chicago press brake. Ram tilt condition causes steel band B to move core of LVDT A. This creates an error signal that, in turn, actuates the pump to correct the tilt condition instantly.

In the case of the press brake, two cylinders are usually provided. A means of balancing the pressure delivered by each cylinder must also be provided. It is in this area that the design of various makes of hydraulic press brake differ the most. Some successful approaches to pressure balancing will be discussed later.

The basic requirements for good press-brake design were set down in Chapter 1. The additional requirements of good hydraulic press-brake design include the following factors that should be considered in specifying equipment.

Basic Design

Some hydraulic press-brake builders buy a number of components from outside sources. This has led, in some cases, to a "Christmas Tree" approach, with motor, pumps, manifolds and piping mounted haphazardly on the brake frame. This approach creates uncomfortable working conditions and a generally awkward clutter. It also is inefficient and can present a hazard. The pump, motor, reservoir and other components of a hydraulic press brake should be compactly mounted. They should be placed where they are readily accessible for maintenance.

Mighty (2500-ton) Steelweld hydraulic press has full tonnage pressing speed of $9\frac{1}{2}$ ipm, rapid advance and high return speed up to 41 ipm.

Motors and Pumps

Design of modern hydraulic motor and pump systems is such that they are virtually foolproof. Their selection will not usually fall within the province of the press-brake buyer as they are customarily supplied as a unit by the builder.

Most modern press brakes use vane-type pumps of high efficiency. In view of the enormous demands on a hydraulic press-brake motor and pump, quality should never be sacrificed for a minor initial saving even if a choice is available. A desirable feature of some units is the enclosure of wearing parts in a cartridge. The cartridge can be removed and replaced without disengaging the pump.

Reservoir and Filter System

The reservoir should be baffled to reduce the surge of fluid that takes place during cycling of the press brake. It should be large enough to promote cooling of the hydraulic fluid, even if this requires taking up useful space. Use of an oil cooler as an alternative to a large tank is effective. A useful feature that can usually be specified, if it is not provided as standard

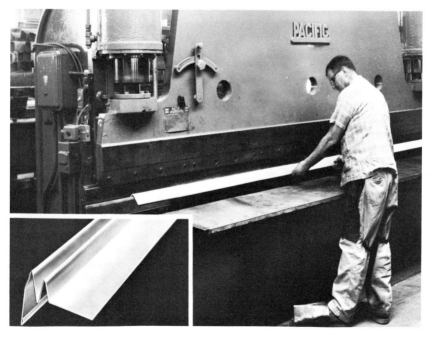

Accuracy of control on hydraulic press brake suits it to repetitive work. On this job it reduces man-hours by 65% on steel flashing work.

Sensitive touch of hydraulic circuit on this big Niagara machine makes possible bending of difficult structurals with excellent accuracy.

equipment, is a temperature gage. Very often such a gage will make it possible to anticipate trouble before it can become serious.

It is essential that the hydraulic fluid be kept as clean as possible. The reservoir breathers should be equipped with filters. Filters should be provided on both the intake and output side of the pump. Where they are not provided on the original equipment, they can be easily and inexpensively installed. The hydraulic fluid itself should be specified with regard to the job to be performed. Fluid of the wrong type or viscosity can cause expensive damage to the equipment.

Manifolds and Piping

Since friction within a tube or pipe varies with its length and straightness, the well designed hydraulic circuit is made up of short, straight sections. Bends and elbows are potential trouble spots.

The better designs use a manifold block that combines valves and piping in one assembly with the shortest possible run to the cylinders themselves. The manifold should be designed so that there is a minimum of high-pressure pipe with as few connections as possible. A design in which the valves can be removed from the manifold without disconnecting fluid lines is desirable.

Counterbalances

Almost without exception modern hydraulic press brakes are equipped with a counterbalance valve. This valve makes it possible to idle the pump between strokes, thus extending pump life. The counterbalance should not depend on positive pump pressure. It should be designed so that the ram will remain in the position at which it comes to rest if the pump fails. This is usually accomplished by using soft packings that retain oil in the cylinders under the conditions described. Since the best of systems will invariably be subject to minute leakage, the ram should be blocked up if the machine is to remain idle for any considerable length of time.

Cylinders

Cylinders on a modern press brake are usually honed or otherwise finished throughout their length. The rods and pistons should be ground. If packings are replaced when necessary in accordance with manufacturers' instructions, service or maintenance on the cylinders should never be required. Highly efficient sealing can also be obtained with multiple metallic piston rings.

Leveling Controls

There are many ways by which the pressure exerted by the two cylinders of a modern hydraulic press brake can be regulated...a function that is obviously essential to successful operation of the equipment. The more successful methods are as follows:

Servo-electric—a ram tilt condition creates a low-voltage electric signal. This signal is amplified and fed back to either of two variable delivery pumps. Advantage—continuous correction of ram level to within .001 in.

Electronic—uses an electronic sensing device of extreme accuracy to maintain ram level. Advantage—simplicity and compactness.

Proportioning valve—uses a proportioning valve to monitor flow of fluid to cylinders and eliminates hunting and wavering. Advantage—continuous correction of level to high accuracy, extreme simplicity.

Limit switch—uses steel-tape sensor that actuates two high-sensitivity limit switches, through spring-loaded cams. Advantage—accuracy and virtual elimination of control maintenance.

Steel tape—uses a steel tape enclosed in dust-proof tube to transmit data to a special level control unit. Advantage—simplicity, accuracy.

Limit switch-relay—uses six limit switches and two relays in combination to maintain ram level. Advantage—simplicity of maintenance.

Hydraulic-mechanical—uses two oscillating rotary-type cylinders of different sizes mounted to a common shaft at the top of the brake frame. The shaft is mechanically linked to the ram through eccentrics on each end of the shaft. When hydraulic energy is applied to the cylinders, the pressure delivered to the ram is always equal. Maximum rotation of the cylinders is 270 deg in any one direction. When a shorter stroke is used, the rotation decreases, thus increasing the potential strokes per min. This type of drive control has as yet only been applied to relatively small press brakes.

The foregoing illustrate successful solutions to the problem of leveling the ram. In each case, the advantages cited are not the sole advantages of the equipment. It is not within the scope of this chapter to recommend specific systems. All the control systems mentioned have given excellent service in hydraulic-press-brake operation and recommendation to that extent is implicit. All are provided with means for deliberately tilting the ram when this is required by the nature of the work to be done.

Hydraulic Maintenance

The hydraulic system on a press brake or shear—or any machine tool, for that matter—can be a source of endless problems. Leakage is probably the most common complaint.

"We've tried everything. Ran the packings up hard. Put in a new filter. Now we just keep a barrel of sawdust around to soak up the oil on the floor." That comment is heard all too frequently. The fact is that if the hydraulic system is of good design—most are—and a good preventive maintenance schedule is followed, hydraulic equipment need never be a source of problems. Proper selection of the hydraulic oil is the most important consideration. Very often, hydraulic oil is bought on a price basis without regard to manufacturers' recommendations.

The oil in the hydraulic system of a press brake or shear is subjected to severe agitation at each pass through the pumps. It is under high pressure. It is apt to be contaminated with water because of condensation within the reservoir. Entrapped air can cause foaming. Any of these factors can lead to rapid oxidation. Foaming, especially, can lead to erratic operation of the brake or shear. There are a number of commercial hydraulic oils on the market, many of which contain additives in the form of inhibitors to prevent precisely those conditions that lead to hydraulic failure or poor performance.

The viscosity of the hydraulic oil is also critical. It must be high enough to prevent metal-to-metal contact, yet it must be low enough so that the pump does not have to labor. The pump manufacturers invariably specify the correct viscosity for their product. Their specifications should be rigidly adhered to. Most plants try to keep the number of lubricants, cutting fluids, oils, etc., to a minimum. If this means that anything less than the best and most suitable hydraulic oil is used, it is the falsest sort of economy.

Chapter 5

BRAKE FORMING THEORY AND PRACTICE

This chapter sets up a check list for planning a job in terms of the many variables that are always present in any bending operation. It discusses air bending vs bottoming, bend allowances, blank sizes, and other factors important to the press-brake user in the production of formed parts.

The bending of a piece of metal seems to be a simple operation. It is actually quite complex. On page 45 is a drawing of a bent piece of metal showing its molecules greatly enlarged. It demonstrates that the inner surface is under compression while the outer surface is in tension.

If this fact is clearly understood, it becomes easier to understand why springback and elongation occur and why there are a great many limiting factors in the brake forming of metal.

The bending process is further complicated by a number of variables— some predictable, some not. These include the temper and thickness of the material, the surface finish of the dies, the nature of the bend, the consistency of the stock thickness, the grain direction and many others.

How, then do you go about planning a job? The first step should be to establish as many positive factors as possible. This can be done by using a mental check list containing the following questions:

1. Do I have adequate bending capacity?

2. Can I air-bend the job or should I use bottoming dies?

3. Will the material bend in the desired manner without splitting or fracturing? What allowance should I make for springback?

4. What blank lengths should be used?

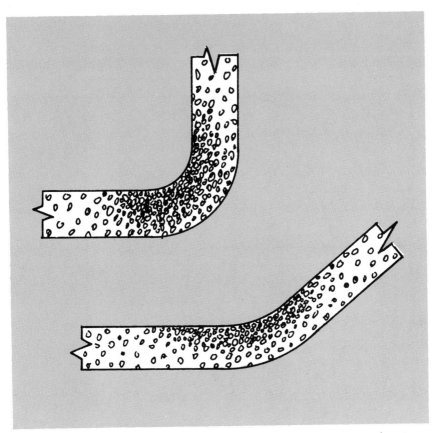

Exaggerated for clarity, drawing shows how molecules are compressed at inner radius; in tension at outside of bend. Neutral axis lies between.

The answers to the first two questions are straightforward. They can be determined positively. The answers to the third and fourth questions are more difficult to arrive at and are seldom completely definitive.

Press-Brake Capacity

The pressure required to bend a piece of metal is a direct function of the hardness and thickness of the metal and the width of the female die across which it is bent. A simple analogy is to picture a 1-in. pine board, 6 ft in length. If you place the extreme ends of the board on two cement blocks, the weight of a small boy will deflect the board. If you place the cement blocks a foot apart near the center of the board, it would take a great deal more weight to deflect it.

The relative position of the cement blocks is equivalent to the width of the female die opening on a press brake die. If you use a 2-in.-thick

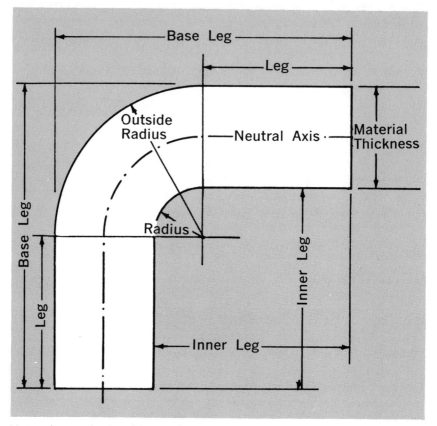

Nomenclature of a bend in metal is quite complex. Familiarity with major terms will help in understanding the discussion on bending theory.

board instead of the thinner board, you also set up a demand for more weight or pressure.

Press-brake capacities are established from the tonnage required to air-bend a 90-deg angle in mild steel—60,000-psi tensile strength—over a female die having an opening at the top exactly eight times the stock thickness. Under these conditions, the inner radius of the bend will be approximately $\frac{5}{32}$ of the die opening, regardless of the thickness of the stock.

The tonnage-requirement chart in Chapter 1 indicates that a pressure of slightly more than 15 tons per foot would be needed to form mild steel, $\frac{1}{4}$-in. thick, over a female die 2 in. wide. If design permitted, only 11.4 tons per ft would be needed to make a bend over a female die $2\frac{1}{2}$ in. wide. The radius of the bend, however, would still be $\frac{5}{32}$ of the die opening—in this case, 0.39 in. as against the 0.31-in. inside radius that would result using the 2-in. (eight × metal thickness) female dies.

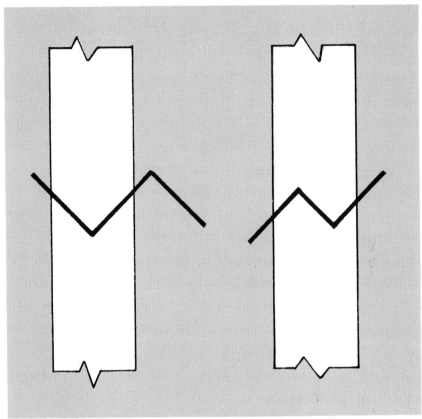

Simple Vee die at left can make an accurate offset bend if the operator is skill-ful. But offset die at right will outproduce it by some 500%.

There is a limit to the width of the female die that can be used even when the radius tolerance is generous. If the female die opening is too large, the lack of concentration of force will cause the metal to bulge along the outer radius.

The rule of eight should not be used when bending stock thicker than $\frac{1}{2}$ in. or with the high-tensile-strength materials. In these cases the die opening should be from 10 to 12 times wider than stock thickness to prevent fracturing. This also reduces tonnage requirements. Remember that the tonnage chart is based on mild steel having a tensile strength of 60,000 psi. Based on the use of the rule of eight, a material with a tensile strength of 90,000 psi would require a tonnage 50% higher than that shown in the chart.

Higher pressures are also needed when the female die is less than 8 × metal thickness. By reducing the width of the female die, a bend with

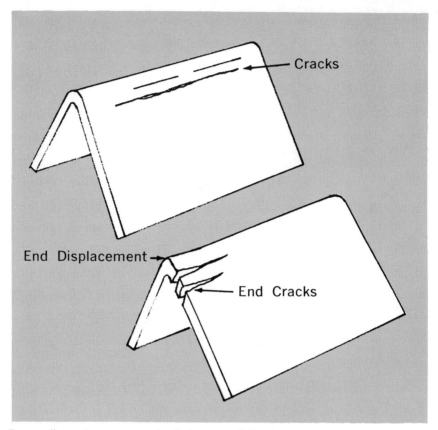

Too small a radius causes major failure parallel to bend direction. Minor cracking can sometimes be corrected by more careful techniques.

a very small radius can be produced. When this is done, however, tonnage
requirements rise sharply and there is danger of exceeding the rated bending capacity of the press brake. Tonnage requirements to bend a piece of
$\frac{1}{4}$-in. mild steel are almost doubled, for example, if the female die width
is reduced from the standard 2 in. to $1\frac{1}{4}$ in.

For bending materials such as aluminum and soft brass, tonnage requirements are greatly reduced since they are in part a function of the
tensile strength of the material. As a rule of thumb it can be assumed that,
for these materials, tonnage requirements are about half those needed for
mild steel.

Air Bending or Bottoming

By use of tonnage charts or standard formulas, tonnage requirements for
air bending can be accurately estimated. It should be remembered that

this is for air bending of single bends. Estimating the tonnage requirements for channels and other multiple bends is not a matter of merely adding up the sum of the bends. Pressure needed to form multiple bends may range as high as 16 times greater than the figure obtained by simple addition. Tonnage estimation for the more complex forms will be discussed in a later chapter on press-brake dies.

Air bending has two basic advantages. It is inexpensive—the cost of air bending dies is far less than that of bottoming dies—and it is very flexible. Almost any conceivable angle or combination of angles can be bent with air bending dies. In air bending, the metal is literally bent in air since the punch does not bottom in the female die. Air bending dies are usually designed in such a manner as to reduce any possibility of bottoming such as in the case of forming a 90-deg angle over a set of acute or hatchet dies.

Lack of accuracy in an air bending die is not the fault of the die. The stroke of the press-brake ram can be set to make a perfect bend on a given sheet. The lack of accuracy stems from the fact that no two sheets of metal are apt to have the same grain structure. If you are forming spring steel, you may find some 12 deg of springback in one sheet and perhaps 16 deg in the next sheet. Obviously the accuracy of the bend is reduced.

Where multiple bends are required in volume, air bending may be at a disadvantage. A simple offset will require two bends and considerable handling. An offset bottoming die, designed to make the offset in one stroke, will have a production rate not merely twice as fast but nearly five times as fast.

Air bending should be specified, therefore, only on short to medium runs of such a nature that a high degree of consistency is not required. Air bending dies should also be specified for use with heavy plate where the tonnage requirements for bottoming and coining would be prohibitive.

Bottoming dies are inherently more accurate than air bending dies. The punch and die are made to the desired angle. Clearance between punch and die is usually just equal to metal thickness. Minor variations in stock thickness are ironed out when the punch bottoms in the die.

Where small radii are required, the clearance between punch and die can be reduced to something less than metal thickness. This causes some coining to take place at the bottom of the stroke. The stock is compressed and springback is reduced. This procedure should be used with extreme care. Tonnage requirements increase drastically and there is the hazard of spoiling the work, at best, or wrecking the press brake at worst.

Far less risky is the technique of using a smaller die opening—as little as five or six times metal thickness rather than the customary eight times. The punch should have the natural radius—$\frac{5}{32}$ × die opening—used in air bending. With this procedure the dies should be bottomed but no coining should take place. Strict attention to punch-and-die clearances is neces-

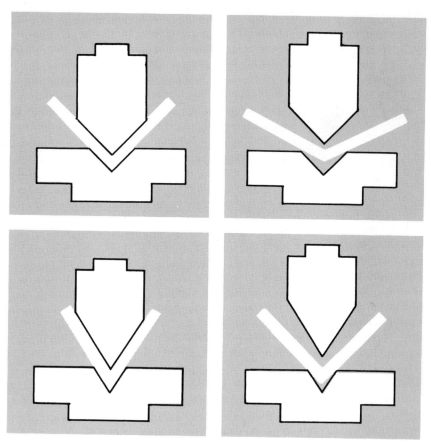

Springback is a major problem in bending. Usual—and often best—procedure to overcome it is to compensate by overbending. Clockwise from top left, the drawings show this accomplished with simple air-bending die.

sary. Tonnage requirements will be higher than those needed with conventional die opening width but less than those encountered when setting the stock by coining.

A third method for making bends of great accuracy with radii less than metal thickness is useful on short runs where special tooling would not be economical. The procedure is to coin the metal at the bottom without regard to the angle of the bend. The work can then be restruck with a male punch from which the forming tip has been removed, using the same female die. The pressure of the ram is directed against the sides of the work, setting the metal so that there is virtually no springback.

When using bottoming or air bending dies it may be necessary to shim the die to reduce deflection. This is more apt to be required with older equipment. Modern press brakes are designed with rigidity commensurate to their tonnage. Where shimming is required, great care should be taken to insure that the press brake is not overloaded. Since deflection will be greatest midway on the ram, shimming should begin at the center.

In general, bottoming dies should be used on all long production runs and on medium to long production runs where accuracy is required.

Far more difficult to determine are the third and fourth basic determinations. Will the material bend in the desired manner without splitting or fracturing? How can the correct blank length be established? Of equal importance is the allowance that should be made for springback.

If a flat strip of spring steel is bent lightly, it will return to its original flatness when released. If a flat piece of ductile copper is bent lightly, it tends to remain bent when it is released. Why is this so? Because the yield point of soft copper is much lower than that of steel. It is essential to note that permanent deformation does not take place until the yield point of the metal has been exceeded. It is also essential to remember that exceeding the yield point does not eliminate springback. Springback always occurs unless plastic flow—uncommon in brake forming—takes place.

This can be understood by remembering that the inner surface of a bent piece of metal is under compression while the outer surface is under tension. Between them lies a plane through the metal where neither compression nor tension exists. This is known as the neutral axis.

As the metal is bent, one part is stretched beyond its yield point—it becomes permanently deformed. The part of the metal that was not permanently deformed tries to return to its original flatness. It exerts stress on the outer surfaces.

Remember, too, that the compressive strength of a metal is far greater than its tensile strength. The compressive yield point is never reached so that the inner surface of the bend will always be trying to return to its original shape even though permanent deformation has taken place on the outer edge.

Veterans of the metal-forming industry will recall that during the war a stretch-forming process was developed for making aluminum aircraft panels. The metal was stretched beyond its yield point while being formed. This essentially created one great neutral plane in which there were no compressive or tensile stresses. Since these did not exist, there was no springback.

In most metal-bending operations, stretch forming would be prohibitively expensive. Springback, therefore, must be dealt with by other means. These are, in order of importance:

1. Overbending.
2. Restriking in a special die after the original bend is made.
3. Coining by bottoming the die in the work. This was discussed in the

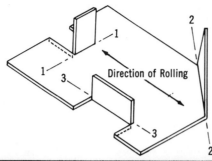

Direction of Rolling

1-1 Bend axis perpendicu-
   lar to direction of
   rolling

2-2 Axis other than per-
   pendicular or parallel
   to rolling direction

3-3 Axis parallel to direction of rolling

Recommended Radii — Forming 90 Deg. Bends*

Alloy	Temper	Nominal Thickness (In.)	Minimum Suitable Radius of Punch — (In.)		
			Bend 1	Bend 2	Bend 3
Copper	Half Hard	0.020	1/32	1/32	3/64
	Extra Hard	0.020	1/64	1/32	1/32
Red Brass, 85 percent	Drawing Anneal	0.005-0.064	Sharp	Sharp	Sharp
	Half Hard	0.020-0.050	Sharp	Sharp	Sharp
	Hard	0.040	1/64	1/32	3/32
	Extra Hard	0.040	1/16	3/32	7/32
	Spring	0.040	1/16	3/16	1/2
Low Brass, 80 percent	Hard	0.020	1/32	3/64	1/16
	Spring	0.020	1/32	1/8	3/16
Cartridge Brass, 70 percent	Half Hard	0.005-0.050	Sharp	Sharp	Sharp
	Hard	0.040	1/64	1/32	3/64
	Extra Hard	0.040	1/32	5/32	7/32
	Spring	0.040	1/16	less than 1/4	less than 1/4
	Extra Spring	0.040	1/8	less than 1/4	less than 1/4
Yellow Brass	Half Hard	0.005-0.090	Sharp	Sharp	Sharp
	Hard	0.040	Sharp	Sharp	1/32
	Extra Hard	0.040	3/64	1/8	3/16
	Spring	0.040	3/64	7/32	less than 1/4
Medium Leaded Brass	Half Hard	0.040	Sharp	Sharp	Sharp
Phosphor Bronze 5 percent	Half Hard	0.020-0.070	Sharp	Sharp	Sharp
	Hard	0.040	1/16	1/16	1/8
	Extra Hard	0.040	1/16	1/8	—
	Spring	0.040	3/32	—	—

Recommended Radii — Forming 90 Deg. Bends*					
Alloy	Temper	Nominal Thickness (In.)	Minimum Suitable Radius of Punch — (In.)		
			Bend 1	Bend 2	Bend 3
Phosphor Bronze, 8 percent	Half Hard	0.005-0.064	Sharp	Sharp	Sharp
	Hard	0.040	1/32	1/8	—
	Extra Hard	0.040	3/64	5/32	less than 1/4
	Spring	0.040	3/32	1/4	1/2
	Extra Spring	0.064	5/32	less than 1/4	less than 1/4
High-Silicon Bronze	Hard	0.020	1/32	1/32	1/16
	Spring	0.020	3/64	3/32	3/16
Nickel Silver 65-18	Half Hard	0.040	—	1/64	1/32
	Hard	0.040	1/16	1/16	1/16
	Extra Hard	0.040	1/8	1/8	3/16
	Spring	0.040	5/32	3/16	7/32
	Extra Spring	0.040	5/32	7/32	1/4
Nickel Silver 55-18	Half Hard	0.040	1/16	1/16	1/16
	Hard	0.040	1/16	1/16	3/32
	Extra Hard	0.040	1/8	5/32	3/16

*Courtesy — Copper & Brass Research Association

first part of this chapter. It is not recommended where overbending will suffice.

Overbending is by all means the simplest and least expensive method for countering springback. The problem is to establish the amount of spring-back allowance that should be made in designing the tool. There are springback tables available, but these are seldom useful other than as guides since so many factors influence the amount of springback that will take place on a given metal.

The following rules will help in calculating springback:

1. The higher the tensile strength of the metal, the greater the allowance that must be made for springback.

2. Springback will decrease proportionately as the angle of the bend decreases.

Trial and error and the application of the foregoing rules are still the most common method for determining springback allowances. Even after the correct allowance has been established, however, variations in stock will reduce the accuracy of the bends. Where this cannot be tolerated, sizing on a second operation die is the only alternative.

Minimum Bend Radii

The minimum bend radii for a given metal, depends on many things including the condition of the tooling, the grain direction and even the condition of the surface of the stock.

There are two approaches that can be used to solve minimum bend problems. If the radius of the bend is fixed and cannot be changed and if cracking occurs:

1. Make sure that the edges of strip or sheet stock are clean cut.

2. Check the grain direction. The bend should always be perpendicular to the grain of the stock. The grain of the stock is parallel to the direction of rolling and is usually marked along the edge in the case of strip stock. Where multiple bends are necessary, compromise so that no one bend is parallel to the grain direction.

3. Try slowing the ram.

4. If these means fail, it will probably be necessary to switch to a more ductile material.

An increase in the bend allowance, where possible, is the most economical way to deal with cracking problems. There is an excellent area for economy here. If the bend radii and the stock to be used are optional, within limits, it may be possible to substitute a higher-tensile, stiffer material in thin gage for a relatively thick, ductile material.

Most steel producers furnish detailed charts showing the minimum bend radii for their products. They are usually readily available. A chart in this chapter shows minimum bend radii for various copper alloys. This chart was provided by the Copper and Brass Research Association.

When charts are not available, a rough rule of thumb can be used to determine a minimum bend radius (R) that will serve as a starting point. For ductile metals R should not be less than 0.02 in. It should be from 0.35 to 0.75 times stock thickness. For hard, high-strength metals, R should not be less than 0.06 in. It should be from 0.55 to 1.5 times stock thickness. The lower numbers given are for thin gages. R increases with the metal thickness.

Blank Length

It has thus far been demonstrated that springback and minimum bend radii are subject to many variables and that they are very difficult to calculate exactly. It would be pleasant to begin the discussion of blank length by stating that here is a simpler matter of arithmetic with no variables and that the correct answer is very easily obtained. Unfortunately this is not the case. Determination of the correct blank length is almost never exact.

As in the case of springback, consideration of all the variables can only lead to an approximation. From that approximation, cut-and-try methods must almost always be used to arrive at the final blank length.

When metal is bent, the compression on the inner surface has less permanent effect than the tension on the outer surface. Consequently the

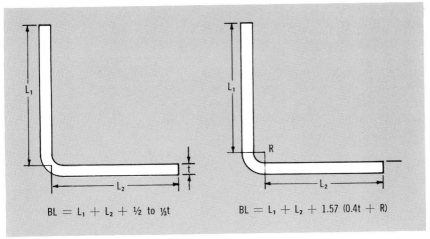

$$BL = L_1 + L_2 + \tfrac{1}{2} \text{ to } \tfrac{1}{3}t$$

$$BL = L_1 + L_2 + 1.57 \ (0.4t + R)$$

Blank-length formula for bends with internal radius less than 3 × T.

metal is slightly elongated. The neutral plane equals the true length of the blank.

A formed shape will always be longer than the blank from which it was bent. The narrower the blank, the greater the elongation. In addition to the many variables already mentioned, lubrication, die finish, the hardness of the dies and many other factors will affect the amount of elongation that takes place.

There are many formulas for calculating blank length. Almost without exception these ignore some of the variables as being impossible to assign a value to. Further, they necessarily assign an arbitrary value to the neutral axis of the metal being bent. There is a wide difference of opinion among experts as to the value that should be assigned.

A simple method for obtaining approximate blank length for reverse 90-deg bends is to add all internal straight sections. To this total, add from $\frac{1}{3}$ to $\frac{1}{2}$ stock thickness for each bend. If the metal is soft and ductile, use the smaller fraction. Increase the stock thickness value directly as the hardness of the metal increases. In using the procedure, you are actually assigning a value to the neutral axis of the metal.

For bends with an internal radius less than three times stock thickness a set of formulas using 0.4 as the value of the neutral axis and an arbitrary factor of 1.57 as a multipler for each bend can be used. Thus, with the drawing as an example, blank length (B) is

$$B = L_1 + L_2 + 1.57 \ (0.4t + R)$$

For two 90-deg bends as shown in sketch, the formula would read:

$$BL = L_1 + L_2 + L_3 + 2 \times 1.57 \times (0.4t + R)$$

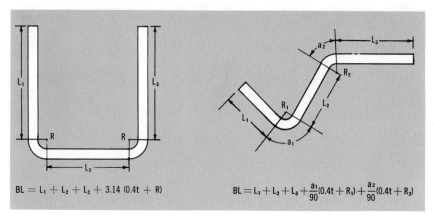

Blank lengths for U-bend (left) and complex multiple bend (right).

For a multiple bend containing both acute and obtuse angles, the formula would read:

$$BL = L_1 + L_2 + L_3 + \frac{a_1}{90} \times (0.4t + R_1) + \frac{a_2}{90}(0.4t + R_2)$$

Several major press-brake builders have made intensive studies of the processes involved in calculating blank length. They have used formulas and actual experiments to determine blank allowances for combinations of angles and lengths in a variety of materials.

It should be realized, however, that the blank length that is correct for a job on one press brake may be off several thousandths of an inch when it is run again unless the dies are set with meticulous care. There seems to be no way of eliminating a final cut-and-try stage in determining correct blank length.

Chapter 6

SMALL AND UNUSUAL PRESS BRAKES

Purchase of a small press brake will eliminate the uneconomical practice of bending small work on a large press brake. A few of the many available models and types are discussed. Some unconventional approaches to press-brake design are described. Many models are illustrated.

Since it would be foolish, economically speaking, to tie up a large press brake on jobs that use only a fraction of its capacity, almost every shop that has large brakes needs one or more small press brakes. Otherwise, the plant finds itself in the position of using a howitzer to bring down a rabbit.

Makers of the truly large equipment recognize this need. Most lines of big brakes are supplemented by smaller or junior lines. As a rule these offer the same features and construction as their larger counterparts, the major difference being that they are scaled down.

One of the common errors in specifying a small press brake is to assume that, because it will be relegated to light work, its construction and design need not be of the quality found in larger models. A second error is failure to specify the features usually incorporated into larger brakes. The thinking here apparently is, "We're only going to be bending flanges on thin stock. We can save a lot of money by buying a used brake without any expensive gadgets on it."

There are several fallacies in this kind of thinking. Invariably, the small press brake that is bought for incidental jobs will be far more useful than was planned for. It will be used on many jobs that were originally sent to the larger equipment. If it has sufficient accuracy and flexibility, it can

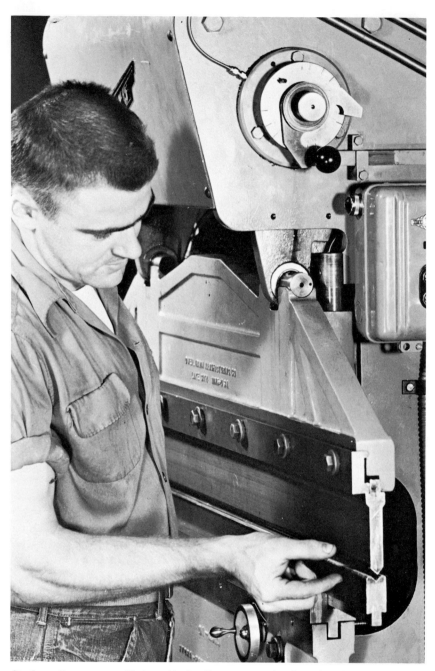

Stroke adjustment on Di-Acro Hydra-Power press brake takes less than 15 sec. Calibrated dial, top right, is guide for dog adjustment.

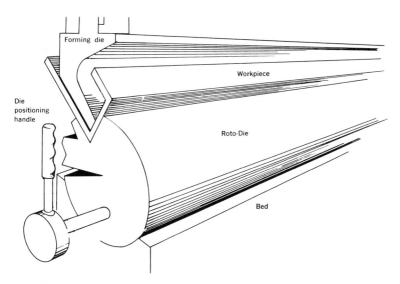

Roto-Die bender by Tu-Bar Corp., has rotating, multi-purpose female dies.   It offers a high order of flexibility to the short-run job shop.

relieve the big brake, save die-changing time, and greatly improve work scheduling and handling conditions.   It can do this in direct ratio to its own capabilities.   If it is slow, inaccurate and of flimsy construction, its usefulness will be greatly limited.

Selection Factors

In setting up specifications for purchase of a small—25-ton and under—press brake, the purpose for which it is being purchased should be carefully defined.   If it is to be an auxiliary to larger equipment, job requirements should be set down.   These include width of stock, gage and production volume.   These determine the most desirable tonnage, width between housings and available speed that should be required in the new brake.   There are many other considerations but these are basic.   Many shops base small brake purchase on minimal jobs, leaving themselves with a very narrow operating range.

An example of this might be the purchase of a brake with a space between housings of 18 in. simply because this width would handle 60% of the odd jobs.   Careful study might indicate that purchase of a brake with a space between housings of 36 in. would handle an additional 20% of the odd jobs.   The point: the potential savings in the extra 20% might greatly exceed the savings possible with the smaller brake.   It is essential to be realistic and not to snatch at a small initial saving in equipment cost.

Selection of a brake as production equipment rather than to relieve larger equipment should be approached with the same degree of realism. Purchase of a brake that barely exceeds minimum requirements as to tonnage, size and speed is seldom economical.

Construction Features

It was pointed out earlier that junior lines of the large equipment makers usually reflect the same principles of good design and construction. There are also a number of manufacturers who make only small and intermediate size presses. These should by no means be overlooked in selecting equipment. Some of them offer design features that are extremely ingenious and useful.

An excellent example is a dieless-type brake, which has two wings in place of the conventional bed. The material to be formed is placed over the wings. It is held in position by an upper forming blade with a knife edge. The blade is mounted in a hydraulically actuated ram. The blade is lowered against the stock and held firmly while the wings, pivoting on a hinge pin that is actually the center of the bend radius, move upward. The motion of the wing is tangent to the bend radius. They form the metal against the stationary forming blade with a wiping action that eliminates whip-up.

This brake can make multiple bends in one operation that would require several steps with conventional equipment. It is available in a number of

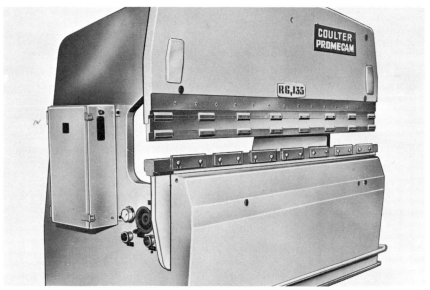

Large—135-ton—version of Promecam hydraulic press brake is also available in 25 through 400-ton models. The bed is the pressing member.

versions with options that make radius bending, box and pan forming, and production bending of large runs feasible.

A number of pneumatic press brakes are available. Usually these are associated with very light work and are designed to take advantage of the inherent low cost of pneumatic actuation. One of the newest models, rated at up to 35 tons in mild steel, differs from the conventional in that the ram is mounted in the bottom of the brake. Pneumatic press brakes have the advantages of low maintenance cost, a cushioned action that makes them very useful on draw work, and quiet operation. They are not generally suited to long-run work, however, except on very light stock.

Hydraulic presses are also available in which the ram is mounted in the bed. These have many of the advantages of the conventional hydraulic design. In addition, there are no obstructions between the housings above the work. The fact that the ram is at bottom-of-stroke when not in use is a safety bonus.

Although it is classified by its maker as a bender, rather than a press brake, another type of equipment with an all-purpose female die merits consideration for short-run work where die change is frequent. Several versions are available. Cut into the lower die of the bender are three die openings and a cross-braking groove. There are two 88-deg openings. One of these is $\frac{3}{8}$ in. wide, the other $\frac{7}{8}$ in. wide. There is an 11-deg opening, $\frac{3}{4}$ in. wide. Finally, there is a 140-deg opening, $\frac{1}{2}$ in. wide, for cross-braking.

Wings, pivoting on a hinge which is the center of the bend radius, shape metal against a stationary forming blade on this die-less Form-All brake.

Automated 24-ton Connecticut press brake equipped with Cooper Weymouth air slide feed. Die has seven stations. Brake operates at 40 spm.

The lower die can be manually rotated to present any desired opening to the upper die. A number of bending combinations are thus immediately available. These advantages can also be obtained by selecting multi-purpose tooling for a standard press brake. (See the chapters on press-brake tooling.)

In another interesting—and very useful—approach to multiple-die equipment design, a four-way female die is provided. All four openings are on the same surface. The die is moved back and forth to center the de-sired opening under the punch. This motion is powered. Standard models have dies for 90-deg bending of 20-ga and lighter stock; 90-deg bendings of 18-ga and heavier stock; for cross braking; and for hemming and flattening up to 20-ga stock. Interchangeable punches are provided for 90-deg bend-ing and cross braking and for hemming and flattening.

A great many options are available for this equipment. With them, virtually any sheet metal work can be done. The ram is hydraulically powered and the depth of stroke controllable. At a 2-in. setting, this machine can make bends at 35 spm. As with other multiple-die equip-ment, the major advantages are flexibility, very rapid changeover from job to job, and versatility essential to job-shop work.

Working area of 15-ton Haggard press brake. Weighing only 1500 lb, this small brake has endless uses in relieving larger equipment for light work.

Regardless of the approach the maker takes to design, certain features are desirable in any small press brake. Among them are:

1. Steel construction. The bed should preferably be welded to the frame rather than bolted. Bed and ram should be massive enough to resist deflection. The ram should be deep for the same reason. Remember that deflection is reduced relative to the cube of the ram depth. Steel gearing is recommended.

2. Drives. Variable speed drives are optional with some makes, standard on others. Where a variable speed drive is an option, its purchase is recommended. The press brake should be provided with a reversing switch readily accessible to the operator. Clutch overload protection should be provided if the clutch is of the air friction type, and clutch and brake should be readily adjustable.

3. Ram adjustment, stroke and speed. These are in part a function of the work for which the brake is intended but, to insure flexibility adequate to the job shop, certain specifications should be met. For a 25-ton mechanical press brake, for example, the ram stroke should be at least $1\frac{1}{2}$ in. and preferably 2 in. Ram adjustment should be at least 1 in. Die space over the bed with ram and adjustment down should be at least 1 in. Die space over a standard die block should be at least 6 in. Such a press brake

with a variable speed range of 20 to 40 spm is truly all-purpose equipment capable of handling almost any job within its tonnage and width capacities.

Most builders offer only mechanical press brakes in the smaller sizes. One builder, however, has developed a combination hydraulic-mechanical press brake that has many of the advantages of both basic principles.

The system uses a reciprocating rotary hydraulic cylinder to drive the ram through a mechanical linkage. The cylinder can be adjusted to stop or reverse at any point desired without changing bed and die settings. Full power is constantly applied. Using this unique system, a speed of 80 spm on a $\frac{1}{4}$-in. opening has been reached. At the full ram opening of $1\frac{1}{2}$ in., this press brake, in a 12-ton version, is rated at 32 spm.

Many other valuable developments have been made in the field of small to medium press-brake design. One builder has developed a brake in which the ram is stationary while the bed does the actual forming. Presently in the design stage is a mechanical press brake in which both ram and bed are driven. Improvements are constantly being made.

Standards and Options

What is standard on one make of press brake is often an option on another brake. To further confuse this issue, what is an option on one builder's 12-ton model may be standard on his 24-ton brake. For that reason, it would be difficult to make specific equipment recommendations here. For the job shop intending to purchase equipment rated in the 25-ton area, however, consideration should be given to these features, whether optional or standard.

Front-operated back gage. Will recover its cost in operator time saved in making adjustments.

Positive lubrication. One-point, positive lubrication will reduce maintenance and possible damage due to negligence.

Power ram adjustment. Should be considered when die changes are frequent or when tipping of the ram for fadeout work is common. Clearly legible ram position indicators should be furnished regardless of whether ram adjustment is manual or powered.

Two-speed operation. Where available as standard or option, this feature will improve safety by eliminating whip-up.

Chapter 7

PRESS-BRAKE DIE SELECTION

A great many factors enter into the proper selection of a press-brake die. A wide range of materials and finishes is available. This chapter points out the factors that govern the choice between inexpensive carbon steel dies and the more costly alloy tooling.

Two factors are common in selecting all press-brake dies, regardless of whether they are standard or special. They are: (1) material and (2) finish. Because errors in specifying either die material or die finish can be costly, these apparently simple factors should be thoroughly studied before a purchase order is issued. Die material and finish, of course, should be considered as functions of the material to be formed, the ultimate surface finish required and the number of formed pieces that will be required.

We can express this in exaggerated terms by stating that few buyers would be foolish enough to specify a chrome-plated die with special tool-steel inserts to form a roof gutter of light-gage aluminum in which the surface finish of the completed part is completely noncritical and of which only a few hundred are made.

Nor would many buyers specify an inexpensive annealed high-carbon steel die with a plain finish to form a half-million parts from stainless steel in which any surface scratching could not be tolerated. Within these exaggerated parameters, however, there is considerable room for error — expensive error. Most commonly this error consists of underestimating die requirements. This is unfortunate because the ultimate cost of under-estimation may extend to thousands of pieces that have to be reworked or rejected.

Overestimating die requirements may cost a few extra dollars initially but this kind of error is not reflected in the work. Referring back to an

Automatic grinder being used to finish a Verson press-brake die. Grinding is demanding job. Improper grinding could result in later heat-checking.

earlier example, you could make aluminum roof gutters with an unnecessarily fine surface finish. This demonstrates a point: if the job requirements are such that a specific die material and finish are not clearly indicated, it is always wise to upgrade rather than downgrade estimated requirements.

There are just four basic die materials used for press-brake forming. Manufacturers of press-brake tooling supply them under a variety of brand names. It should be noted that in all cases the steel used for press-brake dies is special to the extent that great care is taken with carbon content, grain size control and other physicals.

High-Carbon Steel

This is the most widely used press-brake die material. It is supplied with a carbon content ranging from 0.60 to 0.90%. It is a tough, fine-grained steel with good machining characteristics. In the annealed condition it has a hardness of about 205–225 Bhn. This material can readily be heat treated to 267–295 Bhn. The major advantages of this material are its low initial cost and the fact that it is the least expensive of all die materials to machine on the type of equipment usually found in the job shop.

Through-hardened to the range of 245–250 Bhn, this material will give good service on light to medium service applications up to and including the lighter gages of stainless steel. Hardened to the 265–295 Bhn range, the material has good shock resisting qualities and can be used on jobs requiring increased tensile strength as in forming work where the cross sectional area is limited.

Dies of this material can be bought in plain versions. Milling, drilling and other machining can be done in the customer's shop and the die then hardened to about 350 Bhn. This provides excellent wear characteristics but it should be remembered that further reworking or recutting cannot be done without annealing.

Alloy Steel

At a cost about 20% higher than high-carbon steel dies, most makers furnish a high-carbon, high-chrome analysis with an extremely fine grain size and a hardness ranging from 270 to 300 Bhn. This is equivalent to Rc 25–28. This material has much greater tensile strength than high-carbon steel. It use is clearly indicated for severe bending operations, for gooseneck dies where a clearance beyond the centerline is required, and for channel forming dies used to form the more difficult materials such as the heavier gages of steel, stainless and high-tensile materials.

Surface Hardening

For maximum wear qualities a case-hardened normalized medium-carbon steel should be specified. Either flame or induction hardening is suitable. Most die suppliers use the flame hardening process because of its lower cost. Several offer the more expensive induction hardening process. This provides a hard surface over a soft core. Case depth is usually about 0.032 in. thick with a hardness of Rc 48–52. Such dies can be cut to length with a high-speed-steel band-saw but they cannot be machined in the hardened area except by grinding. This latter factor makes the choice between induction-hardened and alloy steel dies a difficult one. The recommendations of the maker should be sought before pinning down final specifications.

Tool Steel

Where excessive wear on limited surface area occurs, tool steel inserts can be used. On very heavy materials where extreme pressures are exerted, over a corner for example, inserts of the release wedge type should be specified. These serve a dual purpose—they resist wear and they help to release material from points where it has a tendency to stick, such as in a channel-type die.

Tool steel can be inserted as one long section in short dies or as a number of multiple sections depending on the length required. It is especially effective where there may be excessive wear from scraping, as in forming over a fairly sharp edge like that found in a spring-pad curling die. This type of insert greatly increases die cost and should not be specified where there is a less costly alternative.

Die Finishes

Finishing processes commonly applied to press-brake tooling include polishing, grinding, honing and chrome plating. As in almost any other area where a choice exists, there is a cost differential. The finishes listed will, in the order given, improve the surface finish of the formed part. They are also increasingly expensive in that order.

Stock press-brake dies are machine-finished to 60 to 70 mu in. These can then be further polished to a finish of 20–30 mu in. This is the most commonly used finish and it is entirely adequate for most materials.

The simpler forms of press-brake tooling can be ground. By using suitable grit sizes, almost any desired surface finish can be obtained. The cost of this processing is in direct proportion to the length of the tooling required.

Many die suppliers use some variation of the wet blasting process for finishing their product. This is essentially a process in which wet abrasive is blasted through special guns at high velocity. The propelling force is compressed air. First step in the process is blasting with a 325- to 625-mesh abrasive. This blends machining marks and removes burrs. The

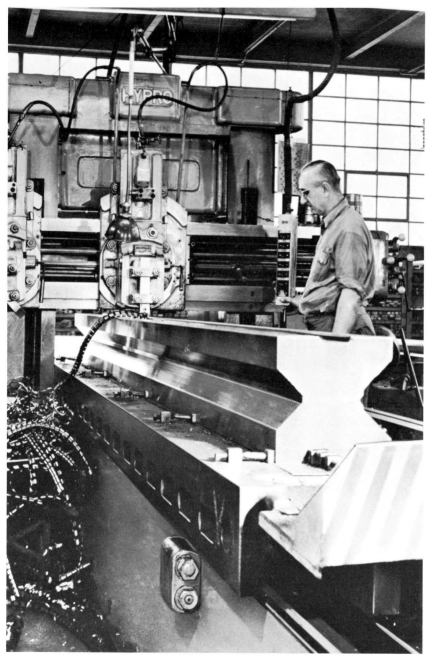

Modern high-speed planer machining a Dreis & Krump 12-in., four-way die. Massive equipment is needed since any deflection would be reflected in the die.

Induction hardening a Dreis & Krump die. Induction hardening is more expensive than flame hardening but it produces the ideal hard-surface, soft-core tool.

die is then blasted with a 625- to 1250-mesh abrasive to remove directional polishing lines, nicks and scratches. The finishing process promotes the free flow of material during subsequent forming operations. A wet blasted surface is excellent for most press-brake forming.

For the forming of polished metal, vinyl-coated steel, painted stock and other material in which any marking cannot be tolerated, chrome-plated dies are recommended. Prior to chrome plating, the die is polished. The chrome plating tends to blend out any microscopic defects in the polished die so that the resulting surface is extremely smooth. Because of the extra cost of chrome plating—about 30 to 40% higher than that of a similar liquid-honed die—chrome should not be specified unless there is a clear need for it.

Mounting Press-Brake Dies

The press brake can only be as accurate as the original setup. A few extra minutes spent in making sure that the setup is correct will always be a good investment. The manual furnished with the press brake has explicit setup instructions. These should be followed exactly. In the lack of a manual, the following procedure can be used for most mechanical press brakes.

The ram should be at bottom of stroke for the initial setup. Some press brakes have markings on the eccentric and the connecting rod. These align when the ram is at bottom, much in the manner that an automobile is timed by lining up a mark on the flywheel. Setup time can be reduced, if the brake has no marking, by painting a stripe bisecting eccentric and connecting rod when the ram is down and using it thereafter as a guide.

For most bending work, a filler block should be used to fill in the necessarily large gap between ram and bed. Filler blocks or dieholders, are

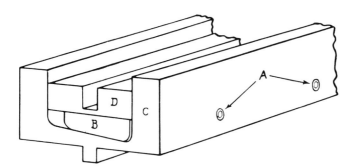

Special bolster for Cyril Bath press brakes contains a sliding wedge (B) which can be driven forward by adjusting screws (A) to raise top plate (D) at desired points. This arrangement eliminates need for shimming.

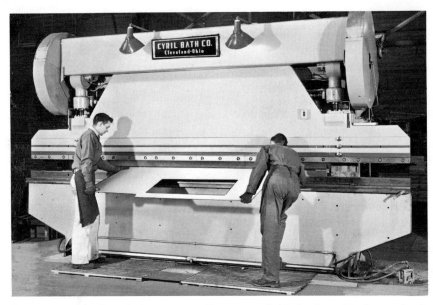

Extensions on both ends increase working area of this brake. Die is standard. Built-in deflection device disengages the clutch when an overload is imposed. This is standard feature on Cyril Bath press brakes.

available in four basic types. They serve several purposes. In filling the die space, they make it necessary to adjust the ram upward for standard dies. This is desirable because it reduces the unsupported section of the adjusting screws. They make it possible to adjust the deflection of the press brake to match different loads. This will be discussed in the section on adjustment.

With the filler block in place but not clamped tight, the lower die should be placed in position and tightened. It is important that the working section of the die to be used is centered between the frames as nearly as possible. Other things being equal, it is always desirable to work at the center of the press brake rather than to one side. The filler block should now be centered so that the Vee of the die is centered over the die tongue slot in the bed and the bolts should be tightened.

Next step is to bring the ram down so that only enough space remains to slide the upper die into place and clamp it tightly enough to prevent it from falling out. The ram should now be backed off and the filler block bolts loosened so that the lower die can be realigned. The ram should now be brought down firmly and the upper die bolts tightened securely.

The ram should now be adjusted upward to the required metal thickness clearance at the die angles. Alignment of the filler block should be adjusted so that the clearances are identical when the filler block is tightened down. Clearance at both ends of the die should be checked, preferably with feeler gages.

Even though the dies are now aligned, it may be impossible to make acceptable bends in long work. All press brakes will deflect to some degree under load. To make a press brake massive enough so that it would not deflect would be economically impractical.

To overcome deflection, a crown is usually machined into the filler block. It is designed so that it precisely balances off the deflection of the press brake under full load so that a perfectly straight bend can be made. Since a perfectly straight bend will be produced only under a specific load, shimming of the die or other measures must be taken to compensate for deflection during overloads or underloads. For overloads, the die should be shimmed at the center. For underloads, obviously, shimming of the ends will be necessary.

Some press brake builders have taken an alternative approach. A crown is machined into the bed and ram of the press brake to compensate for deflection under full load. This eliminates shimming and is a recommended feature in cases where it can be anticipated that the press brake will normally be used for long runs at near full load capacity.

A compensating bolster is available on Cyril Bath press brakes. This patented device has a long sliding wedge mounted in a housing that otherwise resembles a conventional die holder. A top plate over the wedge has a groove that accepts standard die tongues. A series of adjusting screws located at the front of the bolster can be turned to drive the wedge in at selected points, thus creating the same effect as a shim in raising the die. Variations for deflection, material variation, poor die condition, etc., can be made very rapidly. More expensive than conventional Cyril Bath bolsters, their selection can often be justified where long work, in a variety of material thickness, is the rule.

Chapter 8

BASIC PRESS-BRAKE DIES

This chapter discusses the various types of air-bending die, pointing out where they are most commonly used. Selection factors governing gooseneck dies, off-set dies, and hemming and seaming dies are included with recommendations as to their proper use. Tonnage needs are also discussed.

Air-bending dies are widely used in forming sheet metal. A good operator can produce highly accurate work by air bending although speed is sacrificed. In addition to versatility, air-bending dies have the additional merit that they reduce tonnage requirements. There are two major classes of air-bending die; acute dies and 90-deg dies.

Acute Dies

Acute dies of 30 deg are used chiefly in the making of an acute bend prior to closing in flattening dies to form a hem of either the tight hem, tear-drop hem, or open hem type. Any bend under 90 deg is an acute bend

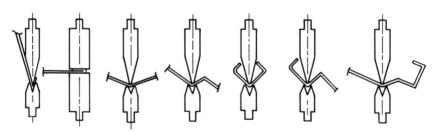

Fig 1—Acute dies can be used to air bend many shapes. Acute bends are usually made prior to using flattening dies in closing various hems.

and tooling, (Fig. 1) can be provided to form the angle required accurately and economically.

By the use of the air-bend method, an acute die set can be used to form angles from 180 deg down to 30 deg simply by adjustment of the press-brake ram until the punch enters the proper depth into the female die to produce the angle desired.

Many shapes can be formed by air-forming with acute dies. The limiting factors are punch width, die width and die height, in case of offsets.

As is the case with most air-formed bends, the finished bend may have a bow from end to end because of variations in material thickness, hard and soft spots in material, and deflection in the bed of the press brake.

The selection of the proper die for material thickness to be formed is based on the formula that the die opening be equal to eight times metal thickness for best results. This applies to mild steel.

90-Deg Air-Bending Dies

The most common tooling used in forming of sheet metal is the 90-deg air-bending die, Fig. 2.

This type of die set is employed to keep forming tonnage low so that a

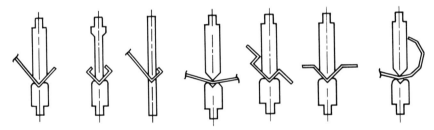

Fig 2—90-deg air-bending dies are the tooling most widely used to form sheet metal. Air bending reduces press-brake tonnage requirements.

lighter tonnage press brake can be used.

The tooling can be used to form 90-deg bends, offset bend channels, and, by adjusting the depth of ram travel into female die, air bends greater than 90 deg.

In forming offsets on 90-deg dies, the limiting factor on the size of offset is that the offset cannot be smaller than one-half the width of the female die. The height of the female die will control the length of the flange that can be produced on the smaller offsets.

In forming channels on 90-deg dies, the width of the punch will control the width of channel and length of flange on the channel

In air-forming angles greater than 90 deg, the finished bend may have a bow from end to end because of the same variations explained under acute dies. Part of the bow may be removed by shimming the die into a crowned condition.

The selection of proper die opening for material thickness is based on the formula that the die opening be eight times the metal thickness through $\frac{1}{2}$-in. material. Above $\frac{1}{2}$-in. material, the opening should be equal to ten times the metal thickness.

Gooseneck Punches

Most tooling manufacturers stock standard gooseneck punches, Fig 3 adequate for forming various material thicknesses.

If changes are made on this type of punch to provide a greater cut-back beyond the centerline to increase the length of a return flange, the work capacity of the punch will be reduced.

Standard gooseneck capacities range from 22-gage material to $\frac{3}{16}$-in. thick material.

A thickness of nose of $\frac{1}{4}$ in. and a depth of return of $\frac{7}{16}$ in. are normal for 22 gage. Nose width should be $1\frac{1}{4}$ in. and depth of return $1\frac{3}{8}$ in. for $\frac{3}{16}$-in. material. There are many sizes between these two.

For heavier materials, the gooseneck-type punches are specially made to

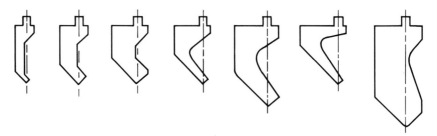

Fig 3—Gooseneck punches are usually stocked in a number of standard configurations. Maker's advice should be sought before they are ordered.

handle specific jobs. If extra clearances are required, these are made in sets with an angle of 30 to 60 deg for light gages to 50 to 40 deg on heavier material thickness.

The customer should check with the manufacturer to see that the gooseneck punch he intends to purchase will satisfactorily perform its specific job.

Offset Dies

Offset dies, Fig. 4(a), and 4(b) are used to form two 90-deg bends at one stroke of the press brake.

Manufacturers have standard sizes that range from $\frac{1}{8}$ to 1 in. by $\frac{1}{8}$-in. increments. Standard offsets are usually based on forming material up to

18-gage mild steel. Other materials and heavier gages of mild steel can be formed, depending on the size of offset and press-brake capacity. Offsets are normally based on the dimension from inside of material to outside of material.

Clearances can be provided on the die set for the forming of hat-shaped channels in two operations on the offset dies.

Offset dies can also be provided to form acute angles, open angles, and

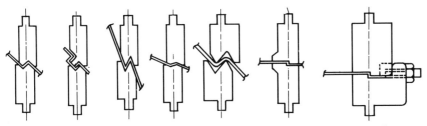

Fig 4(a)—Offset dies are efficient means of making more than one bend in one stroke of the press. These are some of the simpler forms.

to provide clearance for air forming on heavy-gage sheet metal or plate. Air bending will not give the accuracy or form that can be expected with bottoming dies.

The relation between depth of offset and metal thickness affects the accuracy of the 90-deg bends; good results are generally obtained by bottoming on material up to $\frac{1}{8}$-in. mild steel if depth of offset is six to eight times material thickness.

Pressure required for offsetting is approximately five times that needed for a single 90-deg bend in the same material.

For very shallow offsets, there are special offset dies that form the sheet

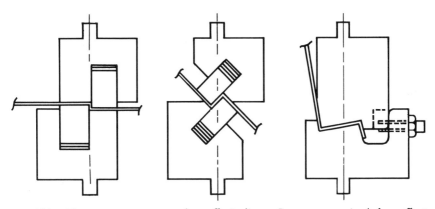

Fig 4(b)—These are more complex offset dies. Pressure required for offset bending is about five times that needed for a single 90-deg bend.

flat. One type is similar to a flattening die with an offset cut into it. This type of die set has no provision for preventing the dies from spreading; therefore, the panel formed on them will have a bow in the length of the offset. The other type has a heel on the back to prevent spreading; this will produce a straight bend. This type of tooling requires pressure approximately eight to ten times that required for a single 90-deg bend in materials to be formed.

For very deep offsets in light stock, there are offset dies with tipped angles to help eliminate some of the distortion that would occur if formed in standard-type offset dies.

Hemming Dies

Hems are used on many sheet metal products. They are used for tucking under the raw edges of material, for stiffening the edges of panels, and as part of a simple lock seam.

There are many ways of making hems. In all cases, however, two operations are required: the first to make an acute angle of approximately 30 deg and the second to close the hem to the desired shape. Tear-drop and open hems require about four times the pressure that would be required to form on a Vee die with an opening equal to eight times metal thickness. The crushed hem requires about seven times this amount of pressure on an average and may run as high as 40 times the pressure that would be required to form on a Vee die with the standard eight times metal-thickness opening.

The simplest form of tooling for forming hems consists of an acute die and a flattening die. These may be combined. The desired type of the three basic hems can be formed by setting the ram to produce the required opening between the surfaces of the dies. This group of dies requires two

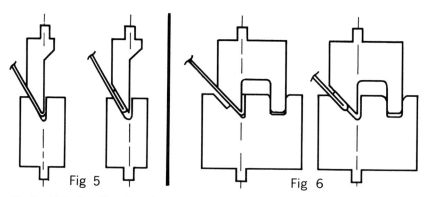

Fig 5—This die will make a hem in two operations and one handling. The leader on the hemming die in Fig 6 helps make a straighter bend.

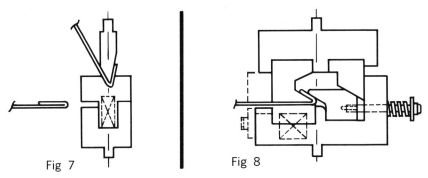

Fig 7

Fig 8

Fig 7—This is a three-high hemming die. Fig 8 is a cam-action die. It's the most efficient and productive of the hemming dies.

operations and two handlings on long sheets or two operations and one handling on short sheets where both die sets can be set in the press brake at the same time.

The preceding section of this chapter discussed the simplest method of producing hems. Additional methods include use of the dies shown in Figs 5, 6, 7, and 8.

The die set shown in Fig 5 will produce a hem in two operations with one handling. The acute shape is first formed. The sheet is then tilted and held high enough on the angle for the punch to clear the flange previously formed and the hem is closed.

Because of side thrust, this operation produces considerable strain on the gibs. A slight bow or camber in the length of the sheet can be expected. Material heavier than 18 gage should not be hemmed in this manner.

The die set in Fig 6 produces a seam in much the same way. The leader on the back of the die makes it possible to bend heavier material and produce a straighter bend. The notch facilitates locating the material and speeds up the operation.

A three-high hemming die set is shown in Fig 7. The acute bend is formed in the top section and closed in the lower. By using an adjustable block in the lower section, tear-drop, open and closed hems can be made.

The most efficient method of making hems is shown in Fig 8. This is a cam-action die. The sheet moves in a straight plane. There is no whip-up and the ram does not have to be slowed during forming. This die set can be equipped with a wider insert on the upper die (dotted lines) and an adjustable angle on the front shoulder. This makes it possible to shim the die to produce the three basic hems. Although this type is the most expensive of the hemming dies shown, it will pay for itself in reduced labor.

There are a number of commercial variations of these basic die types. They should be thoroughly evaluated in terms of productive efficiency prior to making the selection.

Seaming Dies

Many types of seam are used in the metalworking industry. Because of their importance, a number of the most popular types and the tooling required to make them are shown in drawings.

Fig 9(a)—This is a simple lock seam used to join light-gage sheet-metal parts together. The first operation is to form an acute shape on the edge of the two sheets that are to be joined. These are then interlocked and closed in the second die shown for a total of three operations. The seam can be further secured by spotwelding or sheet-metal screws.

Fig 9(b)—The seam shown here is a Pittsburgh lock. The simple tooling shown is for forming the lock if two dies can be set in the press side by

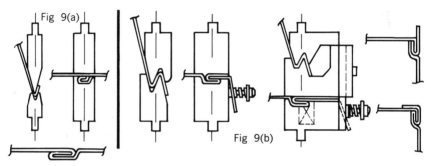

Fig 9(a)—Three operations make the simple lock seam at left. Fig 9(b) is die for making the familiar Pittsburgh lock. Text tells how it's done.

side. The method of operation is to form an acute offset on the first die set, then close in the second. If the part is too long, a three-high die set, as shown, can form the two operations in one handling. In either case, a 90-deg bend would have to be formed on the mating edge of sheet as a separate operation. Turning down of the projecting edge can be done manually with a mallet over a stake or bar to complete the lock or a power tool can be used.

Fig 9(c)—Shown here is the method of forming a double seam, or Acme

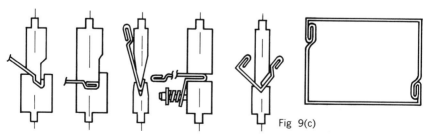

Fig 9(c)—Method for forming a double seam or Acme lock. After pieces are slipped together they can be staked or fastened with sheet metal screws.

lock. First operation is to form an acute with an offset. This is then closed in the second die set and completes one edge. An acute angle is then formed on the opposite edge of sheet. This is then closed in the proper die set. The sheet then has 90-deg angle bends formed into it. This is done on the edges of two parts that are then slid together and secured by staking or with sheet-metal screws.

Fig 9(d)—This standing seam is formed complete in one die set. The panel is then turned over and inserted into the front edge of the die set that is used for the closing operation. This die set completes the standing seam in two operations. The seam is slipped over a 90-deg bend previously formed with a standard 90-deg die set of proper capacity to complete the seam.

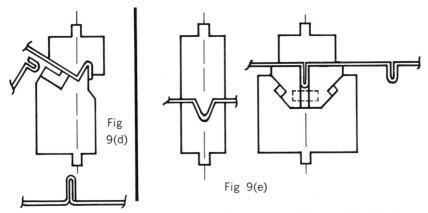

Fig 9(d)—Standing seam is formed complete in one die set. Double flange standing seam is made in the two die sets in Fig 9(e).

Fig 9(e)—This double-flange standing seam is used primarily to give stiffness to a sheet and to provide one-half of a standing seam to join two or three sheets together. An acute bend and two obtuse bends are formed in the first die set, then closed in the second die set. There are some limitations in the use of these dies. In forming a series of seams in a sheet the press must be long enough to pass the sheet between the housings of the press brake. Also, to permit lifting of the shape clear of the lower die, the press must have a stroke at least two and a quarter times the depth of seam.

Fig 9(f)—The connecting cleat is used for connecting two flat sheets with open hems. If the return flanges on the open hem are of proper length, the cleat can be used as a drive cleat. It takes four operations to form the cleat. First operation—form acute bend, second operation—close to shape in lower section of die, third operation—form acute operation, fourth operation—close. Because of weakness in the punch section,

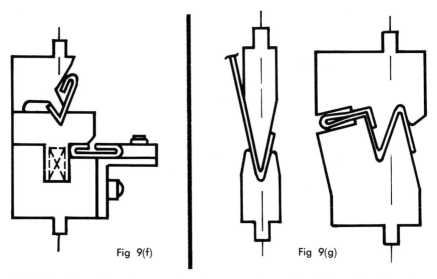

Fig 9(f)—Connecting cleat requires four operations. Die in Fig 9(g) also takes four strokes; makes a seam with great strength and rigidity.

the cleat should be a minimum of $\frac{3}{4}$ in. wide with a minimum gap of $\frac{3}{16}$ in. in order to have some strength in the gooseneck section of the punch.

Fig 9(g)—The "S" cleat offers great strength and rigidity for sections of large trunk and branch lines in commercial warm-air heating and ventilating systems. First operation is to form an acute bend. This is then closed

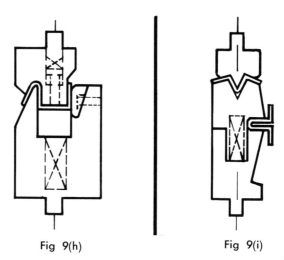

Fig 9(h)—Government lock cup clip can be made in one operation. The "T" cleat in Fig 9(i) is made in a three-tier die set. (See text.)

in the front of the second die set. The sheet is turned end for end and the 90-deg bend and acute-angle offset formed in one stroke. This is then closed on the front of the die set, making a total of four strokes to form cleat.

Fig 9(h)—This shape, the government lock cup clip, should be formed in one operation, using dies similar to those shown. The pressure pad in the lower die will keep the bottom of the channel flat; the pin in the upper die is to force part from the punch.

Fig 9(i)—The "T" cleat can be used to clinch two 90-deg edge flanges together or as an individual stiffening rib. Form two 90-deg bends and acute bend in the upper section of three-tier die set, then close in the lower section. Operator will have to be careful in the closing operation to keep his fingers from between the two closing edges.

Chapter 9

FOUR-WAY DIES AND SPECIAL TOOLING

This chapter discusses a number of approaches to curling, channel forming, and other complex bending problems. These dies are actually standards with many makers, but their purpose is specialized. The work-horse four-way dies are also described. Drawing illustrates the functions of the tooling.

The previous chapter discussed air-bending dies. These are essentially all-purpose tooling since they can be used to make virtually all bends between 30 and 180 deg. Another approach to all-purpose tooling is the use of dies that can be used for two, three and even more operations. Usually they are confined to a class of work—curling, channel forming, etc. An exception is the four-way die—a very useful piece of tooling with exceptional versatility.

Four-Way Dies

The familiar four-way dies come in many shapes and sizes. Although they are primarily designed for 90-deg bending, they can be furnished with one acute opening and also one flat side for closing. They can be furnished with various special bends.

Four-way dies should run the full length of the press brake so that a suitable lifting link can be hung from the end of the ram to slide over the pin. The purpose, of course, is to make it easy to rotate the die when a change in die opening is wanted.

A minor limitation to this kind of die is that offset forming is controlled by the size of the block. If one of the die openings is used much more than

Lifting links make rotation of four-way dies easy. This one is for full-length dies. Attachments are also available for shorter multiple die units.

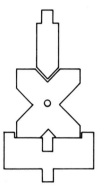

Fig 1(a)—Typical four-way installation of punch, four-way die, and holder.

Fig 1(b)—Four-way die has an acute shape, 90-deg openings, and plain side for flattening.

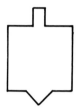   

Fig 1(c)—Special type of four-way die has punches that can be used to form a number of special shapes, including curls and flying-vee shapes. Variations on this theme are virtually unlimited and, often, highly profitable.

the others, it will wear more rapidly and may require recutting at the loss of the other openings.

Fig 1(a) shows a typical four-way die installation of punch, four-way die and holder. In Fig 1(b), the die has an acute shape, 90-deg openings, and a plain side for flattening. Fig 1(c) shows a special type of four-way die with punches that can be used to form radius, flying-vee shapes, curls and special edge-bending.

This represents only a minor fraction of the possibilities available with four-way dies. Very often, the economy of this multipurpose tooling will warrant purchase of specials.

Curling Dies

A great many types of curling dies are available. They all have the same function—to produce a false or wired edge. There are two major types; usually referred to as off-center and on-center.

The tooling shown in Fig 2(a) will form a curl over a rod in three operations. In the first operation, an acute shape is formed. In the second operation, the work is closed partially around a rod at the front of the die set. The final operation is finish-form, curling around the rod at the rear station of the die.

The tooling shown in Fig 2(b) will form a curl with or without a rod. The first and second operations are done in the first die set. The final operation is closing to size in the second die set.

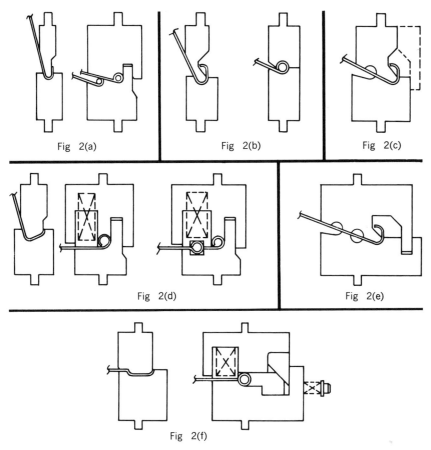

Fig 2(a)   Fig 2(b)   Fig 2(c)

Fig 2(d)   Fig 2(e)

Fig 2(f)

Curling dies come in a number of forms. The ones shown here are typical. Purpose is usually to produce a false or wired edge in sheet-metal work.

The die set in Fig 2(c) will form a curl in three operations in one die set. The first and second operations are done in the rear section of the die, the final operation in the front section. The limitation on this type of curl is the diameter of the curl in relation to metal thickness, which can require excessive weakening of the gooseneck nose. If the part is too heavy or if it exceeds 6 ft in length, a leader (dotted lines) may be required to keep the dies from spreading apart.

The dies shown in Fig 2(d) will form an off-center curl in two operations or an on-center curl by addition of a groove in the pad and lower die as shown in the second closing die set.

The die shown in Fig 2(e) will form an off-center curl in three operations or an on-center in four operations as was done in the die shown in Fig 2(c).

On-center curls in lighter gages of metal can be made in two operations with the dies shown in Fig 2(f). First operation is preforming of the two radii. Next, a cam-operated die set rolls the curls to finish shape while the sheet remains in the horizontal plane.

Remember that radius bends will require about three to four times the tonnage needed to form material over the proper 90-deg die. Closing the bends will require five times the tonnage needed to form material over the proper 90-deg dies with material up to 14 gage. Over this thickness, the tonnage will jump almost 100%. The preform bends shown in Figs 3(d) and 3(f) require about five times the tonnage required to form material over 90-deg dies with the proper opening.

U-Form Dies

As with curling dies, there are any number of U-forming dies available as standards and specials. A few of them are shown in Fig 3(a) through 3(g). Again as with curling dies, radius forming requires from three to four times the pressure required to form material over 90-deg dies with the proper opening.

Fig 3(a) shows a rocker-type die used to form a U-bend on the edge of a sheet. This type is suited only to light-gage material and small-diameter bends. It forms the U-bend in one operation. The tooling shown in Fig 3(b) is a rotator-type die set suitable for U-bends in 12- to 14-gage material and lighter. Up to 2-in. diam bends can be made with this tooling. It requires only one operation.

The rocker-insert type die in Fig 3(c) is designed to overbend the material. This compensates for springback so that the legs are straight and parallel when they are released from the die. One operation is required.

Two operations are needed to form a U-bend with the die set shown in Fig 3(d). Since this has a gooseneck-type forming punch, there is a limit

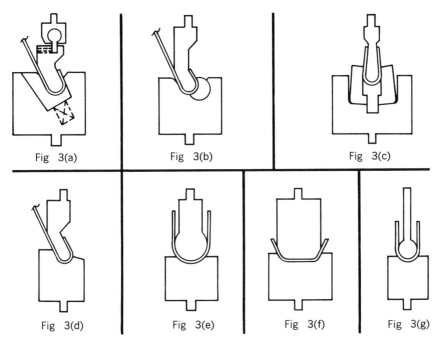

Fig 3(a)   Fig 3(b)   Fig 3(c)

Fig 3(d)   Fig 3(e)   Fig 3(f)   Fig 3(g)

U-forming dies, standard and special, are accurate and fast method of making radius bends. Pressure required is much higher than in air bending.

to the size of the bend and the thickness of the metal—otherwise, the gooseneck portion might be excessively weakened.

Fig 3(e) requires three operations to make a U-bend. Dies of this type are designed for use with larger diameters and heavier materials. The die sets shown in Figs 3(f) and 3(g) will make U-bends in two operations. With appropriate gages mounted on the dies, they can make very accurate bends. They can be used on light, medium, and heavy-gage material and in large and small diameters, depending only on the die space and tonnage of the press brake.

Channel Dies

Channel dies reduce the number of forming operations required to make channel shapes but they require considerable pressure. Their design requires many considerations. This can be best explained by steps. Fig 4(a) is the very simplest type of channel die. With this type of die, the formed part may stick in the lower die or on the punch. The bottom of the channel will have either a concave or a convex shape and will not be truly flat. The punch and die could be curved to compensate for this but it would not be a consistent correction. By adding a release insert to the lower die, as in

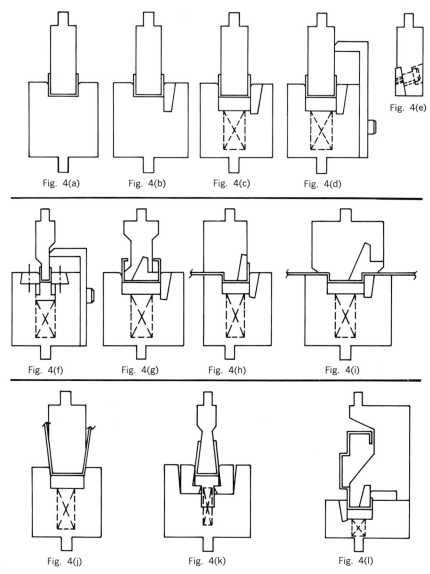

Fig. 4(a)   Fig. 4(b)   Fig. 4(c)   Fig. 4(d)

Fig. 4(e)

Fig. 4(f)   Fig. 4(g)   Fig. 4(h)   Fig. 4(i)

Fig. 4(j)   Fig. 4(k)   Fig. 4(l)

Channel dies are also very productive but considerable care must be taken in their design. "Tricks of the trade" improve their performance.

Fig 4(b), sticking of the part in the lower part of the die can be overcome. The part can still stick to the punch, however, and it will still have a curved bottom.

By adding a spring pad in the lower die, Fig 4(c), the sticking problem is eliminated and the part will have a flat bottom. It can still stick to the punch. In Fig 4(d), a solid-type stripper has been added that will remove the part from the punch. There are some limitations to this solution. They include the leg length, thickness of the material, and accuracy of the corner bends.

In Fig 4(e), a release insert has been added to the punch. By this means, and with a die with a spring pad and release insert, the part will have a flat bottom and will not stick to the die or the punch. This is the most efficient and desirable type of channel die set.

Some additional types of channel die are shown as follows:

Fig 4(f)—a channel die for use on narrow channels.

Fig 4(g)—a channel die set for wider channels and special shapes.

Fig 4(h)—a die set for forming a channel on the edge of a sheet.

Fig 4(i)—a die set for forming hat channels or stiffening ribs.

Fig 4(j)—an open-angle channel die set.

Fig 4(k)—a channel die set with rocker inserts for use when compensation for springback is necessary.

Fig 4(l)—a special die set used with hat channel die set or offset dies in forming metal door frames.

Dies sets a, b, c, d, e, f, g, j, k and l require about five times the tonnage needed to form over the proper opening 90-deg die. Die set h requires seven times that tonnage and die set i requires ten times that tonnage.

Box-Forming Dies

A high punch and die of the type commonly used in forming boxes is illustrated in Fig 5. The actual depth of box that can be formed is a function of the ram width, the punch height and the die space available in the press brake. This can be established as a formula.

Where ph is punch height

bd is box depth

w is ram width

then bd (max) $= 0.7\left(\text{ph} - \dfrac{\text{w}}{2}\right)$

and ph (min) $= \dfrac{\text{bd}}{0.7} - \dfrac{\text{w}}{2}$

This type of punch should have a hook tongue so that the punch can be cut, shaped and matched. In this manner, various sections can be combined as needed to make boxes of various lengths and widths.

A practical combination would include sections of (1 inches) 1, $1\frac{1}{4}$, $1\frac{1}{2}$, $1\frac{3}{4}$, 2, 3, 4, 5, 6, 7, 8, 12 and 24. This would provide a very useful range of working sizes. Sections shorter than 1 in. are not recommended. The sections listed would permit forming of boxes of most sizes up to 50 in. square.

The hook tongue should be designed so that it provides easy assembly of sections without the risk of falling out of the ram clamping bar. This punch could also be furnished as a high goose-neck punch for use in forming boxes with return bends on the top edge. There are applications where a horned and mitered section at each end would be required to clear an in-turned flange at the top of the box.

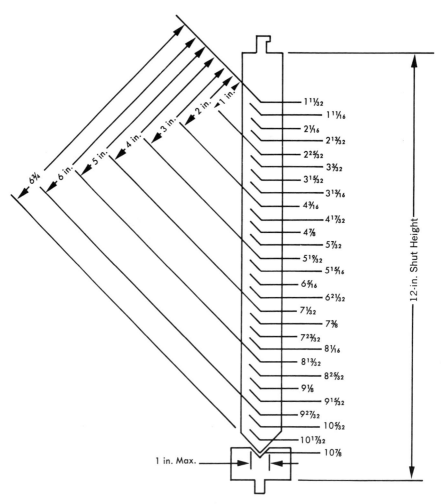

Box-forming dies typically have a high punch and die. Depth of box that can be formed depends on ram width, punch height, and available die space.

Chapter 10

BENDING WITH "SOFT" TOOLING

On many press-brake bending jobs, solid blocks of urethane and other elastomers can be used to replace conventional machined dies. The savings, of course, are great. This chapter discusses some possibilities and limitations of the "soft die" techniques and points out where they can be used.

One of the most interesting and useful developments in press-brake tooling is the use of rubber-like materials that replace female dies. The urethanes are the most advanced of these materials. Their value lies in their ability to transmit kinetic energy. Urethane, unlike rubber, is virtually incompressible. If a blank is forced into a block of the material by a punch, the energy of the punch is transmitted almost evenly in all directions. If the material is confined by a retainer, the energy is reflected back by the retainer walls since it has no place else to go. It will force the blank against the punch with uniform pressure.

If a finger is pushed into a black of rubber, it tends to make a cone-shaped depression. The urethane more nearly resembles a cup full of steel shot. Push a finger into such a cup and the steel shot will tend to conform to the shape of the finger. This is the principle on which soft tooling operates. A number of sources for this kind of tooling are available.

The advantages of urethane tooling include the following:

1. Low tool cost. Urethane is usually used in block form with minor relief to direct metal flow. It can be machined with ordinary shop equipment.

2. Minimum setup time.

3. Extreme flexibility.

4. Non-marring of surface finishes.

Material here is 20-ga steel. Die is a standard Kaufman model. Cost of the urethane tool was $393.30, plus $65.00 for the punch. Die is reusable.

This is experimental job. On production run, a urethane die pad would be used to extend life of die by protecting it against sharp die edges.

5. Excellent part definition when properly used.
6. Production of sharper bends or smaller radii than with metal dies.
7. Sharp reduction of springback and wrinkling.
8. No compensation needed for variation in stock thicknesses.

Until recently, urethane dies in press brakes were specified to meet one or more of the preceding requirements. The present trend is to accept urethane as a standard die material and its use as a standard practice even where its "special" benefits are not a factor.

Production quantity is not necessarily a pertinent factor. The specific application determines whether urethane tooling is suited to short, medium or long runs. For example, an application where extreme deflection of the elastomer occurs, such as in heavy-gage deep U-forming, will cause greater die-pad fatigue than a common V-forming operation, using lighter metals, without deep punch penetration. In the latter case, good definition can be obtained after many thousands of strokes without appreciable wear on the die pad.

Elastomers such as DuPont Elastomer Chemical Dept's Adiprene L combine great abrasion resistance, high modulus, excellent oil and oxidation resistance and cut resistance.

There can be compounded in such a manner as to strengthen any of these properties to meet special tooling requirements. Thus, when a

Job is complete. Note simplicity of punch even though finished part has a rather intricate shape. Potential of urethane is barely scratched.

standard, general-purpose urethane die pad is not applicable, a special can be formulated at no great cost.

Urethane Die Pads

Standardized tooling setups for press brakes are composed of various sizes and grades of urethane die pads used interchangeably in metal die retainers. An air channel in the retainer beneath the pad extends service life by evenly distributing internal compressive strain throughout the pad. This stress-relief channel also helps control deflection of the elastomer for deeper penetration and more efficient forming. The metal retainer must have sufficient strength to withstand the pressures transmitted by the elastomer under load. A thin urethane wear pad used as a buffer between the die pad and the metal blank will prevent cutting of the pad during long runs.

When the punch depresses the metal blank into the flat die pad, the urethane temporarily deforms. It produces high bottom and side pressures uniformly throughout the pad. These pressures form the metal into the precise shape of the punch, without marring the surface. Springback is virtually eliminated. The paid then returns to its original dimensions,

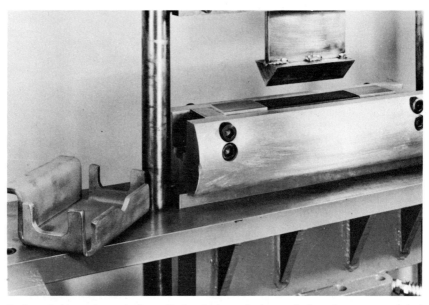

Train switch part is made from ¼-in. thick steel plate. Only a few were needed so the blank was flame cut. Die pad was cut slightly longer than the finished part and placed in retainer. Gaps were filled with metal inserts to enclose die-pad ends. Tool cost was $158.00 including urethane.

ready for the next stroke. Precise stroke adjustment is not needed.

The most important factor in designing a tooling setup is maintaining the required pressures in the die pad for a particular range of metal gages and shapes to be formed. Punch shape, depth of penetration, and the characteristics of both pad and metal are the variables that must be considered.

Generally, punch penetration should be limited to one-third of the combined pad thickness plus the air space under the pad. For most work, pad hardness should be in the 85–95 durometer range. Since most urethane grades can accommodate a range of applications, there is considerable latitude in selecting die components. When switching from light to heavy gages, it will often be necessary to change to a harder die pad. For maximum service life, over-deflection of the die pads should be avoided.

The machinability of urethane is an advantage in many cases. For instance, since it is often important that punch-die pad lengths be approximately equal, the same die retainer can be used for short and long lengths by simply adjusting the length of the urethane pad. The pad can be saw-cut for shorter punches, or short sections of urethane can be butted together in the retainer for longer punches. They perform like one piece

Under pressure, urethane bulges upward. The inside radius of the formed part was 1/16-in.; much sharper than conventional methods allow. Low production item like this would ordinarily be made by heating it with a torch and bending it over a forming block clamped in a vise.

under compression. When cutting a longer pad to a shorter punch length, the ends must be confined to maintain forming pressure in the pad. This can be done by filling in the vacant end areas with bar stock.

Typical Applications

A number of urethane tooling applications are shown in the text and drawings. Note that many of them can be performed with the same die pad and retainer. For example, a press-brake pad selected for 22-gage-steel will readily form metals up to 12 gage.

Vee Forming

This a common Vee-forming application in a standard retainer. Under compression of the punch and blank, the elastomer bulges slightly above the retainer walls and also into the stress-relief air channel under the pad. A fabricated retainer can also be designed, using deflector bars to form the air channel. This provides stress relief and also properly controls pad deflection. The solid hydraulics principle of this method tends to equalize forces throughout the pad to provide uniform pressure at bottom of stroke.

U-Forming

U-bending jobs can often be done on Vee-bend die setups. Since the accuracy of forming to the punch is governed by the amount of final bot-

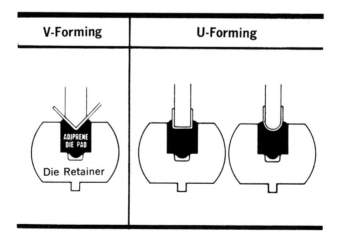

toming, greater pressures are required. To prevent over deflection it is important that the punch width plus metal thickness at bottoming does not exceed a maximum of 60% of die-pad width.

Gooseneck Punches	Box Forming

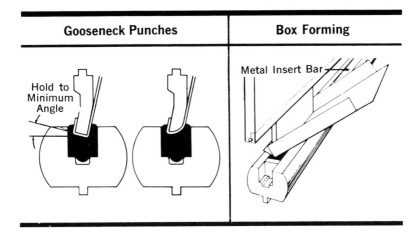

Radius Forming

For radius forming, a wider die pad and retainer are needed. Adjustment of the timing of bottoming pressure may also be needed. The easiest method is to decrease the area of the stress-relief channel beneath the pad by inserting filler bars. This causes the elastomer to deflect upward with greater force, permitting faster bottoming pressure with less punch penetration. This same procedure can be used when forming heavier gages—with the same tooling—than otherwise recommended.

Gooseneck Punches

The drawing shows two typical setups for forming with gooseneck punches. Urethane is somewhat limited in these applications because punch design is restricted to the approximate minimum angle shown. This procedure is suited only to the lighter gages.

Box Forming

In this case; the die has been end-packed with a metal insert bar to maintain pad pressure. The width of the flange and its inside radius must be considered. For mild steel, minimum flange should be four times stock thickness plus the inside radius or $\frac{1}{8}$ in., whichever is greater. With softer metals, shorter flanges can be safely formed. Longer flanges are needed with harder metals.

Side-Pressure Forming

Special side-pressure forming can be accomplished with good definition by using preshaped urethane die pads. A cast or machined-to-shape pad, roughly approximating the shape of the punch, provides adquate pressure

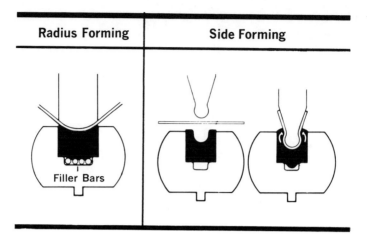

Radius Forming	Side Forming

Filler Bars

at bottoming for these difficult bends. The pad acts as a conventional steel female die until the punch is near bottom; then it deflects and forms the blank entirely around the punch.

Reverse Bend Forming

In using standard urethane dies for reverse bending and for bends needing deeper penetration, it may be necessary to raise the pad to delay timing of the bottoming. This can be done by merely shimming the pad to provide clearance during the second stroke of the reverse bend. This increases side pressure when deeper punch penetration is needed. This procedure limits use of the die to lighter gages since a corresponding loss of pressure occurs at the unrestrained top of the pad.

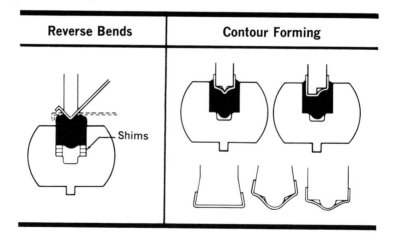

Reverse Bends	Contour Forming

Shims

Contour Forming

Compound bends and contour forming with concave areas present a special problem because urethane cannot deflect into sharp recesses. With lighter gages, sharper definiton is possible.

Wipe-Down Tooling

A recent development is a wipe-down tooling design using high-durometer urethane forming blocks. Instead of a universal female die, the specially formulated urethane acts as a precision, nonmarring forming punch. These machinable forming blocks can be set for minimum clearance since the urethane will deflect slightly to accommodate stock-thickness variations. To replace springs, another standard urethane material functions as a pressure pad during wiping action. This provides hold-down pressure without marring the blanks.

Urethane can also be used as a stripping pad in piercing and templet blanking dies. It can be used for extrusion, bulging, drawing, and embossing dies and roll forming.

Chapter 11

PUNCHING AND PRESSWORKING WITH THE PRESS BRAKE

The long bed of the press brake makes it very useful for many jobs other than bending. This chapter discusses methods for extending the die area by lengthening and widening the bed and the selection and use of adjustable, reusable tooling as an important means of lowering costs.

The press brake is one of the most versatile pieces of metal-forming equipment. It can perforate, punch, trim, blank, notch and draw—and it can do these things economically.

The particularly long, narrow configuration of the press brake and its ability to deliver force in a confined longitudinal area were developed for the forming and bending of plate and sheet metal. These same characteristics make it an ideal means of processing long and bulky shapes that would not fit within the confines of a press. It is equally well suited to pass-along work in which a great many punching units are ganged along the bolster. To make room for the installation of the necessary die sets or punching units, a wider bolster is required than would be used for straight bending.

Ideally, the press brake should be bought for one application only— bending or punching. This is seldom feasible. Most shops want to be able to go from one type of operation to another with a minimum of delay.

Widening the Press Brake

The press brake should be widened or provision for widening should be made at the time of purchase. When this is not possible it can be done

Unusually well-guarded Cyril Bath PT press brake in a pressworking application. Guarding of this nature, unfortunately, isn't always economically feasible.

Rubber cylinders are used for strippers in this application. Punching bolster is a Cincinnati type MN. Most brake builders supply punching bolsters.

in-plant by two methods. Angle brackets can be welded to the ram and bolster or they can be bolted on. Welding is cheaper and provides a more rigid structure. It detracts from the flexibility of the press brake, however. Unless no further need for the press brake on straightforward bending operations is seen, bolted angles should be used, in spite of the greater expense of drilling and tapping the necessary holes.

Press-brake builders furnish equipment already machined to receive angle brackets on the ram and bolster. The slight additional expense should not be a deterrent if there is the remotest chance that a wider bed will be required later.

Press-brake builders usually offer three additional options: (1) The press brake can be supplied with a permanent wide bolster and a removable angle bracket for the ram. (2) It can be supplied with permanent angle brackets on both the ram and bolster—these are integral parts of the machines. (3) Finally, the press brake can be supplied with removable angle brackets on both the ram and bolster. Of these variations, the latter will usually be the best choice. The user should be guided, however, by the maker's recommendations based on the anticipated workload.

The angle brackets themselves can be commercial structural shapes or special fabrications, depending on requirements. They may be interchangeable, but it is usually advisable to use a wider angle extension on the ram to compensate for its narrower width.

Gag blocks can be pulled out to allow punches to ride up in holders. This is a flexible arrangement for punching a variety of straight patterns.

If the installation is done in-plant, standard practice for mounting should be followed. This provides for the use of 1-in. bolts on 8-in. centers. The angles may extend through the housings, but usual practice is to confine them to the nominal length of the ram and bolster. Short lengths can be used to provide for punching and bending work without requiring removal of the angle brackets.

There are practical limits to the amount the bolster and ram can be widened. These limits are a function of the press tonnage and several other factors. They vary with the press brake and the maker but the practical limit, on a large press brake, would be in the area of 36 in.; wider units have been successfully used on specific applications. Where the punching and progressive working of a wide stock will be the rule rather than the exception, particularly in the heavier tonnages, a straight-side press with four-point gibbing should be considered. Several makers offer this kind of equipment, which combines the versatility of the press brake with many of the advantages of a press.

Lengthening the Bed

Bed and ram extensions are available for all press-brake makes. This should be specified as original equipment, if at all possible. Extensions for equipment already in use are difficult to install and never have the rigidity of original equipment. It is almost impossible to use a press brake for

Special die set uses spring plungers for strippers. Job here is punching holes in grain bin side sheets. Load on punches should be closely figured.

horning operations, such as the closing of box ends, without an extension on one end. It will usually pay, on most operations, to order a press brake with extensions on both ends. One end can then be used for horning; the other can be equipped with angle brackets so that an extra operation can often be obtained almost "free."

The policy of buying a shorter press brake than is required and using extensions to make up the difference is a poor one and is not recommended. Deflection of the portions of the ram and bed outside of the housings will be so great that die shimming will be a constant problem. It will more than consume the small savings in initial cost.

An interesting approach to long lengths is the mounting of two press brakes side by side, using them as a unit. Extreme lengths can be formed or punched in this manner without the risk of severe deflection. Applying this theory with widened rams and beds provides what is an effective "assembly line" for metal fabrication.

Draw Cushions and Strippers

When press-type dies are installed in a press brake for drawing operations, cushions are required. Rubber cushions are adequate for light drawing. Air cushions are recommended for deeper draws. If the use of air cushions can be anticipated, a double-plate bed should be specified.

For variable center distances, adjustable-center punching bolsters can accept various die sets; add much to the versatility of the press brake.

This simplifies addition of air cushions. The recommendations of the press-brake maker with respect to cushions should be followed explicitly.

Punching and Notching

There are almost limitless means of tooling a press brake for multiple punching, notching and perforating. The only boundary is the ingenuity and imagination of the user. Press brakes have been tooled to punch several hundred holes, of many configurations, in one stroke—while a second operation was being performed on an extension.

Most press brakes will accept standard narrow dies without additional support. Wider dies can be mounted on angle brackets mounted to the press-brake ram and bed. Standard punching and piercing bolsters are available from most press-brake manufacturers. A typical unit accepts standard punch and die sets and provides for adjustable hole spacing in a straight line. Usually these have guide pins at each end of the bolster to maintain alignment. This kind of equipment is best suited to long-run operations in which case it is enormously productive. For work when the holes are in a straight line but in different patterns, special gag-type punch holders are available. When the pattern changes, the punch can be set so that it rides up in its holder and doesn't punch the metal.

For holes in metal up to 12 ga, simple rubber strippers are usually ade-

Type MN bolster used with double-row special punching units to make holes in railroad car flanges. Job typifies versatility of press-brake punching.

quate. For heavier material, special provision must be made for stripping. One excellent solution is a beam mounted on the inside of the press-brake frame. The beam accepts solid strippers and makes it possible to punch holes anywhere on the sheet. This, of course, is a limitation when the strippers are mounted to the die holder.

Regardless of the punching method, some care must be taken not to overload the press brake. Elsewhere in this chapter is a tonnage chart showing the tonnage requirements for punching holes in a number of sizes and material thicknesses. The total tonnage required should, under no circumstances, exceed two-thirds of the rated capacity of the press brake.

Calculation of the tonnage required to punch a number of holes is a matter of simple addition. For noncircular holes, the outside circumference, divided by three, can be used as a factor. Multiply this figure by 80 to arrive at the correct tonnage requirement. The diameter of a hole multiplied by metal thickness and by the factor of 80 can be used when a tonnage chart is not available or where it does not list the required hole size.

Where the tonnage requirement approaches the limit, the punches should be stepped. By setting half of the punches needed on a level one-half metal thickness higher than the remaining half, the required tonnage will be halved. By setting the punches in three or four levels, the tonnage

Tonnage Requirements for Punching Mild Steel Plate

Thickness of Metal Ga.	In.	Hole Diameter (In.)														
		1/8	3/16	1/4	5/16	3/8	7/16	1/2	9/16	5/8	11/16	3/4	13/16	7/8	15/16	1
20	0.036	0.35	0.53	0.71	0.88	1.1	1.2	1.4	1.6	1.8	1.9	2.1	2.3	2.5	2.7	2.8
18	0.048	0.47	0.71	0.94	1.2	1.4	1.7	1.9	2.1	2.4	2.6	2.8	3.1	3.3	3.5	3.8
16	0.060	0.59	0.89	1.2	1.5	1.8	2.1	2.4	2.7	2.9	3.2	3.5	3.8	4.1	4.4	4.7
14	0.075	0.74	1.1	1.5	1.9	2.2	2.6	2.9	3.3	3.7	4.1	4.4	4.8	5.2	5.5	5.9
12	0.105	1.0	1.6	2.1	2.6	3.1	3.6	4.1	4.7	5.2	5.7	6.2	6.7	7.2	7.7	8.3
11	0.120	1.2	1.8	2.4	3.0	3.5	4.1	4.7	5.3	5.9	6.5	7.1	7.7	8.3	8.8	9.4
10	0.135	—	2.0	2.7	3.3	4.0	4.6	5.3	6.0	6.6	7.3	8.0	8.6	9.3	10.0	10.6
3/16	0.187	—	2.8	3.7	4.6	5.5	6.5	7.4	8.3	9.2	10.2	11.1	12.0	12.9	13.8	14.8
1/4	0.250	—	—	4.9	6.2	7.4	8.6	9.8	11.0	12.3	13.5	14.8	16.0	17.2	18.5	19.7
3/8	0.375	—	—	—	—	11.1	13.0	14.8	16.6	18.5	20.3	22.1	24.0	25.8	27.7	29.5
1/2	0.500	—	—	—	—	—	17.2	19.7	22.1	24.6	27.1	29.5	32.0	34.4	36.9	39.4
5/8	0.625	—	—	—	—	—	—	—	—	30.8	33.8	36.9	40.0	43.0	46.1	49.2
3/4	0.750	—	—	—	—	—	—	—	—	—	—	44.3	48.0	51.7	55.4	59.0

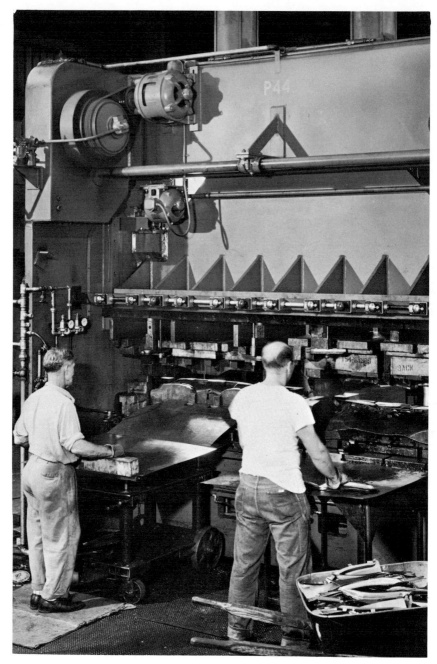

Pressworking on a Cyril Bath press brake. Work is passed through from the front. Elaborate control system insures operator safety at all times.

requirements will be one-third and one-fourth the total punch area, respectively. Where necessary, tonnage requirements can be reduced by grinding a shear on the punch.

For the shop needing a great deal of versatility in punching operations, adjustable and reusable tooling is strongly recommended.

In its simplest form, adjustable and reusable tooling consists of a punch and die, holders for same, and a stripping device—there are several of these. Each unit provides a range of hole sizes. Invariably, punches, dies and other components in a given size and make are completely interchangeable. With unit tooling, most components are interchangeable, regardless of make, in the more commonly used models and sizes.

Various makers provide different means of securing the punches and dies to the holder. The major variation in reusable tooling, however, lies in the manner in which the tooling is attached to the press brake.

Die-Set Types

One of the many useful approaches to reusable tooling uses what is in effect a precision die set. The die set has T-slots, top and bottom, usually running parallel with the front-to-back line of the press. The punch and die holders are slotted to accept two T-bolts. With this arrangement, it is possible to locate the required punches and dies at any point within the die-set area.

The setup is usually made with templets, using pilot and setup plugs to locate the punch and die holders. This kind of tooling is available in standard sizes, in a number of shapes, for holes from $\frac{1}{32}$ to 3 in. in diameter.

Another version in which a die set is used utilizes two drilled and tapped plates rather than T-slotted components. The punch and die retainers are slotted. The tapped hole pattern is such that, depending on retainer width, holes can be spaced on minimum $\frac{3}{4}$-in. centers. More than 30 holes can be punched on a die set with a 9 × 12-in. area.

This equipment is set up with a templet, using pilot plugs. Templets can be made by shop personnel or by the equipment maker. Of much interest is a "starter set" offered by a major supplier; this permits users to inaugurate a reusable tool program at nominal cost.

Still another version of the die-set type of tooling, designed to reduce setup time on work up to $\frac{1}{4}$ in. thick, uses magnetized punch and die holders. These are located to a templet of the needed pattern—magnetism holds them in position while snap rings are installed. The templet is then installed in the die set. An advantage of this system is that setups can be made outside of the press, reducing downtime to a minimum.

Thus far we have been discussing tooling that is incorporated into die sets. The punches and dies in their holders are separately mounted. The greatest single advantage of this type of mounting is that the die area is

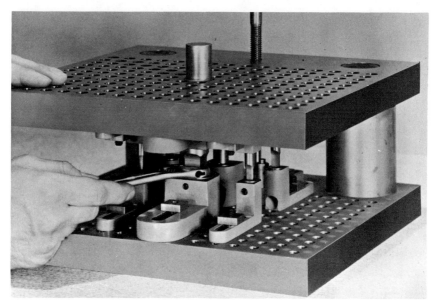

Universal die-set tooling by Di-Acro uses master mounting plates with precision spaced holes. Crank is for setting correct punch height. Stem shows at top.

clear of obstructions. Progressive and feed-through work is feasible and hole centers can be close together. There is no limit to hole patterns or work size other than the width of the die set itself.

Self-Contained Unit Tooling

Self-contained unit tooling consists of a punch and die, integrally mounted in a C-frame. Protruding from the bottom of the frame is a pilot pin. The pin is concentric with the punch and die. The punch is independent of the press-brake ram.

Proper alignment of the punch and die is a function of the holder and is not affected by deflection in the ram, by worn or sloppy gibs, or other deficiencies in the equipment.

The versatility of this kind of equipment is exceptional. Inexpensive horizontal units can be used to punch holes on vertical flanges or extrusions. Specials can be made up to accommodate virtually any configuration. As with die-set mounted tooling, they are not confined to holes and squares. Slots, ovals, irregular shapes and notches of any configuration can be made.

These need not be specials. Potential users should check with the manufacturer before ordering a special. Sometimes a minor modification to a standard unit will make it adequate for the job. These "semi-specials" are relatively inexpensive.

Arrangement of Unipunch tooling for long line of regularly spaced holes. Simple piece of angle iron can be used as feed rail; stock support.

Mounting Unit Tools

There are a number of methods for mounting unit tools in the press brake. Since mounting methods affect accuracy, cost and tool-change time, they should be carefully studied at the time the decision is made to adopt reusable tooling as a basic manufacturing method.

The most accurate method of mounting unit tooling is with a templet. Usually the templet is a piece of $\frac{1}{2}$-in.-thick hot- or cold-rolled steel. The pattern is laid out on the steel. Holes for the pilot pins are drilled and reamed. Holes for the hold-down bolts are drilled and tapped. When extreme accuracy is needed, the templet should be ground and the pilot pin holes should be laid out on a jig borer. (All manner of ingenious transparent models and layout devices are available for templet preparation. Most users will learn to "improvise" with them.)

When the appropriate tools are bolted to the plate, the job is complete and ready to run. When the job is done, the setup can be broken down, the tools stored or reused, and only the templet must be retained for future use. This templet method should only be chosen where high precision, repeat runs or large quantities are involved.

A variation on the templet method uses a universal mounting plate that has a number of tapped holes on close centers in a regular pattern. The templet itself can be quite thin—as thin as 12-gage and under—since it does not have to serve as a mounting plate for the C-frame. The universal

Very fast setup is possible with magnetic-type unit made by Whistler. Units are located and held by snap rings prior to final assembly of tooling.

plate (or a T-slotted plate) serves this purpose. The templet can be punched, rather than drilled or jig-bored. Only the pilot hole locations are critical. Engagement of the pilot, even in thin templet material, is sufficient to prevent it from shifting.

The rear portion of the C-frame, if it overhangs, is shimmed to the thickness of the templet before it is bolted down. When the templet is so large that it interferes with the bolt slots in the tool, clearance holes can be drilled or punched to allow the holddown bolt to be inserted.

This is an extremely fast method that is accurate enough for most jobs. It reduces the cost of templets and the time required to make them since the simplest kind of shop equipment can be used. This method can be considered for production runs so short that unit tooling would not ordinarily be feasible. Some shops go this practice one better—they secure the retainers with double-sided, pressure-sensitive tape.

Press-Brake, Unit-Tooling Accessories

There are several special press-brake accessories that should probably be purchased if long and bulky work or work requiring long lines of holes —cabinets and appliance work are good examples—is to be done.

If the holes are close to a common center line—under 5 in.—purchase of a bed rail is recommended. This attachment mounts on the press-brake bed. It has T-slots in the rear to accommodate the hold-down bolts in the

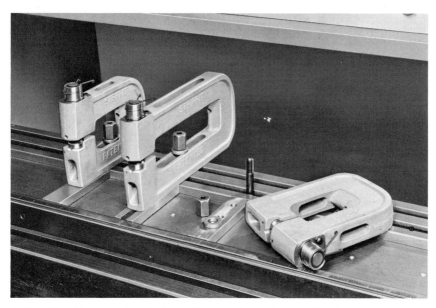

Bed rail in this setup has T-slots. Note scale at front ·of brake. It speeds up alignment of punching units. Infinite variety of patterns can be made.

tooling. It usually has a milled grove, $\frac{1}{2}$ in. deep by about $4\frac{3}{4}$ in. wide, on its front edge. The groove accepts a strip templet into which pilot pin holes in the desired pattern are drilled. Only pilot pin holes are required on the templet.

A variation is to put spacers in the templet groove instead of a long strip templet. Usually these are spring-loaded and can be snapped into place instantly. They have pilot holes in one side and slots in the other. This makes it possible to stagger the tooling to a limited degree.

Another version of the bed rail has a milled groove that will accept the tongue of a standard press-brake die. Purpose? So that conventional press-brake tooling can be installed without removing the bed rail. Also, if a press-brake bending die of the correct height is used, forming can be combined with punching and notching in a progressive operation.

It would be impossible to discuss all the devices that have been developed to take advantage of the inherent advantages of the press-brake, unit-tooling combination. Potential users of the method should write to major suppliers of this kind of equipment, many of whom provide staffs of highly trained field service men.

A hatful of useful accessories are available for use with unit tooling. Which ones should be bought are a function of the work for which the tooling will be used. The point is to know that they are available. Among them are:

Arrangement of Strippit punching and notching units provides for adjustment front-to-back and side-to-side with minimum lost motion.

Feed rails. These are strips resembling angle iron that support the work. They cost very little in view of their usefulness. Magnetic versions are available that do not require bolting down.

Stops. These are available in straight pin, magnetic adjustable and other versions. Function is self explanatory.

Pickup Gages. These are used to locate the work in pass-along operations from previously punched holes. Spring-loaded versions are available.

Locator Pins and Plugs. These are used in layout and setup. They are inexpensive but very useful.

Ram Adapter Plates. These are plates that mount to the ram and protect it against marking from the punch heads. They also protect the punch heads by covering holes and slots in the ram. Most shops make these plates up themselves. Their use is recommended.

Selecting the Equipment

Which type of reusable tooling is best? This decision can only be made in terms of a given type of work. All of it will do what its maker claims it will do. Not only are there differences in tool design, there are differences within types. Most makers of this kind of tooling supply heavy-duty as well as standard versions.

They also offer different means of stripping the work from the punch. For some work, a simpler rubber or urethane stripper is adequate. For heavier work, spring and even spring-hydraulic strippers are required.

Cincinnati press brake is fed by roll hemmer and is fully automated. Production is extremely high. Output is about two miles of roof gutter per hr.

A working knowledge of the potential in reusable tooling can be gained by study of the excellent literature on the subject published by the various makers. Even a brief exposure should convince the reader that is a basic manufacturing process, of great flexibility, that has a tremendous cost-saving potential.

Unit-Tooling Maintenance

Reusable tooling is rugged, accurate and reliable. Unfortunately, some users equate the word "reusable" with "expendable" and treat the tooling accordingly. Remembering that, once the concept is adopted, you will un-doubtedly acquire a valuable inventory, adequate provision should be made for live storage and for storage of a supply of replacement punches, dies and accessories.

Tooling should not be carelessly stored in a bin. When a setup is broken down, all components should be cleaned. The punches and dies should be sharpened, and the complete unit inspected.

The punches and dies should be sharpened well before they start making unacceptable holes. There is a good reason for this—it will extend punch and die life by as much as 300%. One maker suggests that about 15,000 holes can be punched in $\frac{3}{16}$-in. stock and about 25,000 holes in 14-gage stock before the punch needs attention. The number drops off sharply as the thickness increases.

Freshly ground, the punch and the die button have square, sharp edges.

Hydraulic press brake arranged to shear and punch at the same time. Materials is 1-in. thick mild steel. Production rate is 50 pieces per hr.

As the tools wear—the punch wears more rapidly than the die—the edges become rounded and less shiny. The dull look means that the edges have been crystallized. This is the proper time to grind the tool since it will be necessary to remove only a small amount of material—somewhere between 0.015 and 0.020 in.

Caught at this point, a punch can be sharpened many times. Once crystallization starts, the punch requires more tonnage to punch a hole. Crystallization spreads rapidly and it may be necessary to grind off much more metal to produce a good edge. Instead of a number of grinds, you may be able to make only three or four. The punch will also require excessive shimming.

Care in handling of the retainers will assure that they retain the alignment built into them.

Shearing With a Press Brake

Standard and special shearing attachments are available for press brakes. They are not recommended for high-volume production shearing because they do not provide the accuracy or the speed of a power squaring shear. They can be used, if high speed and accuracy are not required, as a substitute for flame cutting and other cutting methods.

Assuming that a brake of adequate capacity is available and that it has a low utilization rate, purchase of a shearing attachment may be justified for intermittent work that would otherwise be contracted outside or flame-cut.

The inaccuracy of shearing on a press brake stems from the nature of the shearing forces. A press brake is designed primarily to exert a purely vertical force. The forces in shearing are not confined to the vertical plane. There is a strong horizontal thrust that tends to force the knives apart.

Holding down the work is also a problem. Part of the design of a good shear is based on developing effective holddown pressure, often by cams and hydraulic cylinders. Most shearing attachments for press brakes have spring-powered holddowns. On heavy material in particular, they may not deliver sufficient force to keep the work from shifting slightly.

Chapter 12

THE DESIGN AND CONSTRUCTION OF SQUARING SHEARS

A squaring shear is almost always used as a working partner to the press brake. Modern shears can produce blanks to near die-cut tolerances. This chapter discusses basic design features of the standard shear and the factors involved in selecting equipment best suited to the type of work.

The modern squaring shear is a precision machine, designed and built to meet severe demands for accuracy and performance. To meet these demands, builders have developed a variety of construction techniques and manufacturing methods. The modern shear can easily hold tolerances to within 0.005 in. if properly used and maintained. No single element is responsible for providing the high performance standards of today's equipment. The potential buyer of a shear has the choice of these approaches to meet his needs:

1. Iron or steel construction.
2. Bolted or welded construction.
3. Mechanical or hydraulic power.
4. Overdrive or underdrive.

Since so many options exist, it would be impossible to attempt to designate the best choice here. Such a choice must depend to a great extent on the nature of the work. It is, however, possible to point out some of the major effects of the various considerations on shear performance.

Choice of a shear will also be strongly influenced by the nature of the operation. Is it a high-volume production type job? Or is it an intermittent-use, short-run application? This is an important consideration. If

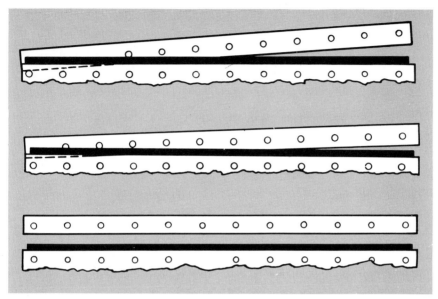

Exaggerated for clarity, the drawing illustrates rake. Adjustable only on hydraulic shears, the amount of rake can effect camber, bowing and twisting.

the shear will be used only for the occasional squaring of light stock, a heavy-duty, production machine may not be required. If the shear will be used for day-in, day-out operation, in a variety of stock thicknesses, selection of a reliable, heavy-duty, more rugged shear is essential.

Shear Materials and Construction

The major makers of shears have taken three approaches to structural design. Shears may be made of rolled steel plate, machined and assembled with keyed and bolted construction. They may be of welded-steel or cast-iron construction.

The various manufacturers advance strong arguments in favor of each of these materials and construction approaches. The dampening characteristics of nodular cast iron, the greater tensile and shear strength of steel, are typical of these arguments. The claims are valid to a degree but they are not so important as other factors. Both materials and all three types of construction are entirely adequate to shear construction, provided that they are properly employed.

Shear material, as used by major makers, is not a major factor in shear accuracy, provided that structural design for that material is sound. It must be remembered that the major load on a shear is a shifting or travelling load and that torsion stresses as well as horizontal and vertical stresses are developed during the cut. To insure accurate shearing, the crosshead

Heavy plate shear has a capacity of 1½ in. in mild steel. Rake angle can be rapidly changed. The equipment can be used for heavy plate or extremely thin metal.

and bed must be rigid enough to resist deflection in more than just the vertical plane and also twisting from torsional stresses.

Rigidity, as applied to shears, is more a function of structural design than of mass or material alone. Also essential are adequate bearing areas and gibbing to guide and confine the ram. The counterbalances must be capable of partially absorbing the severe shock of breakthrough. This is a very important factor in production work.

The other function of the counterbalance is to offset the weight of the ram assembly and to eliminate "floating" by holding the ram snugly against the guides. Spring counterbalances are commonly used on many shears. Air counterbalances are also supplied, particularly on larger shears.

The motor should have an ample power margin. On continuous operations, at relatively high speeds, there is little time for the flywheel to store energy. This places a demand load on the motor. An under-powered shear is not an efficient production tool.

The clutch, whether mechanical or air friction, should be reliable. It should have a good "safety record." Accidental repeats in a shear cannot be tolerated because of the very great risk involved. It should be as maintenance-free as possible.

The equipment supplied by most U. S. manufacturers more than meets these requirements. The exceptions are a handful of imports and a few low-priced models.

Purchase of a shear on the basis of low initial cost is not recommended. The operating economies possible with a quality shear are extremely high. It can shear to close limits so that secondary operations can be reduced. It can shear accurately and it will retain its accuracy under production conditions without the need for constant adjustment. It is designed in such a manner that adjustments can be made rapidly when they are needed. It can operate at its rated speed virtually without downtime. These economies will more than offset the higher initial cost of good equipment.

It should be pointed out that the quality of the metal being cut has a strong bearing on the accuracy of the cut and the quality of the sheared edge. The very finest shear cannot compensate for deficiencies in the quality of the stock being sheared although it is often given the role of "goat."

Hydraulic or Mechanical

In an earlier chapter of this book, the comparative merits of hydraulic and mechanical press brakes were discussed. The same considerations apply to a somewhat lesser degree in making the choice between a mechanical and a hydraulic shear.

Other factors being equal, the mechanical shear is faster than the hydraulic shear. One maker who builds both types rates his mechanical

Heavyweight Niagara shear on a typical job. This is gap or overdrive construction. Picture on opposite page is underdrive model. Note difference in height. Lights under ram are for shearing to scribed line from front of shear.

Underdrive construction improves visibility. Shears of this construction are very successfully applied to line-type operations such as an automated cutoff facility. The low silhouette of the shear is an advantage in such installations.

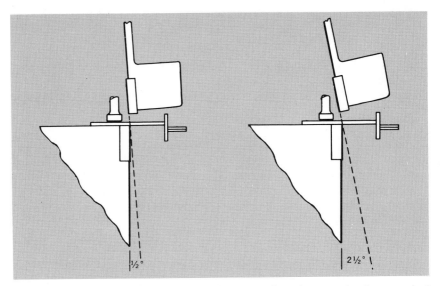

Relief angle must be provided to prevent binding off cut between back gages and knife. Text on page 129 describes several approaches to provision of clearance.

shear with a capacity of 18 ft in $\frac{1}{4}$-in. mild steel at 30 spm. His hydraulic shear, with the same length and capacity, is rated at slightly more than 15 spm. This relationship is fairly typical.

Hydraulic shears are usually associated with heavy-plate fabrication although some new equipment, in the smaller size ranges and with relatively high operating rates, is now available. The major advantages of hydraulic shears is the provision of rake adjustment. By increasing the rake angle, heavier metal thicknesses can be handled and the capacity of the shear extended. It should be remembered that when this is done, with most makes, the knives must overlap at the low end so that the available effective shearing length is reduced.

Hydraulic equipment has the additional advantage that overloading, while possible, does not ordinarily result in structural damage. The choice between hydraulic and mechanical equipment will be strongly influenced by the nature of the work for which it is purchased. In most cases, it will be fairly clear cut.

Overdrive vs Underdrive

Most shears use overdrive construction; that is, the driveshaft, gearing, flywheel and motor are above a throat or gap provided in the side frame members. The major benefit of this construction is that it permits slitting of stock longer than the shear is wide. The stock, of course, passes through the gap between the frames. Underdrive shears are more compact and

Although high speed is a relative term, this shear qualifies under any definition. available in range of capacities, it's commonly operated at speeds to 300 spm.

have a lower silhouette with improved visibility. For straight-through work or incorporation into fabrication lines, this is a very great advantage. The effect of drive type—gap frame or underdrive—has only minimal effect on shearing characteristics.

Rake, Ram Angle and Clearance

Rake is the slope of the knife from side to side. If rake is held to a minimum, camber, twist and bowing of the sheared material or off cut will also be minimal. There is a penalty, of course. As the rake angle decreases, the vertical load on the brake increases. The rake angle is essentially a compromise. It is designed to produce a commercially acceptable off cut at minimum practical loading on the frame, bed and crosshead. Shear builders have established effective rake angles for their product based on the capacity and length of the shear. An average rake for a $\frac{1}{4}$-in., 12-ft shear would be about $\frac{1}{4}$ in. per ft. Rake would increase to approximately $\frac{5}{8}$ in. per ft in a 12-ft shear designed to cut 1-in. steel.

Makers of hydraulic shears almost universally provide rake adjustment. This makes it possible to increase the rake for cutting heavier stock and extends the capacity of the equipment with no penalty in reduced stock-length capacity. This is not generally done on a mechanical shear.

Overdrive or gap-frame mechanical shears do have a vertical adjustment of the crosshead, which provides for raising the crosshead without changing the rake angle. The purpose is to set the knives so that they do not

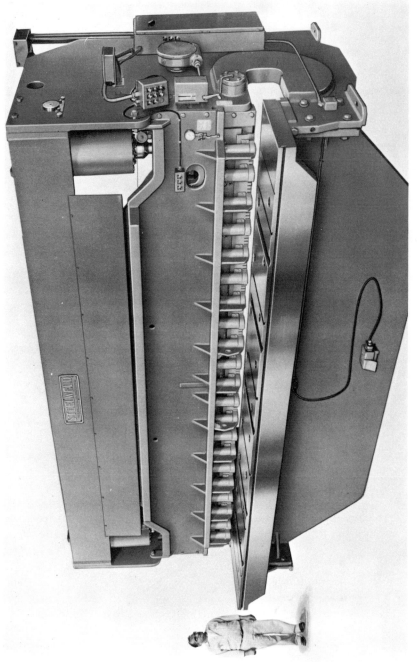

Huge hydraulic shear can make 20-ft cuts in ¾-in. thick mild steel. Single crank adjusts rake and clearance simultaneously.

overlap at the high end so that slitting can be done easily.   As noted, this does not increase or decrease the rake and has no effect on loading.

Another item of construction difference among manufacturers is the manner in which a relief angle is provided between the off cut and the fixed knife. The prime purpose of this relief is to let the off cut fall away without binding between the back gages and the knife. It also has the function of eliminating the possibility of knife interference or rubbing.

Major builders have taken several approaches in providing this clearance. Some of them are:

1. The crosshead descends in a true vertical plane. Provision for mounting the lower knife is such that it is set at a small—about 2-deg—angle relative to the upper knife.

2. The crosshead itself is inclined downward and inward. One maker uses an angle of $\frac{1}{2}$ deg. Most makers use a somewhat larger angle. On the order of $2—2\frac{1}{2}$ deg would be typical.

3. The crosshead is not mounted in vertical slides but is pivoted. It descends in a very small segment of an arc whose point of tangency with the bed is at the cutting point.

All of these approaches have given excellent results.

Knife clearance—the space between the knives—is another controversial area in shear design. Most shear builders have taken the position that, if the rigidity of the shear is adequate, stock up to the capacity of the shear can be cut with the same clearance used to shear stock of minimum thickness.

Other builders feel that the knife clearance should be adjusted for thick-

Self-compensating plunger holddown is one of the many features of Pexto (Peck, Stow & Wilcox Co.) model 10-U-10. Clutch housing is removable on this series.

Wysong shear with optional light beam and back gage calibrated to 0.001 in. Note high visibility of back-gage readout. This one is enroute to customer.

ness being cut to obtain a clean cut with minimum burr. They provide a means of adjusting the knife rapidly. Whether or not to adjust for various stock thicknesses is a function of economics, the size of the run, and, to a degree, personal preference.

Use of a fixed minimum setting eliminates the possibility of shearing thin stock with too much clearance. When this happens, the sheared piece can wedge between the knives, causing severe jamming.

Correct initial setting and accurate grinding of the knives are essential to good shear performance. These points will be discussed in detail in a later chapter.

High-Speed and Special Shears

High speed in a shear or any other mechanical device is a relative term. There are two classes of high-speed shears. One is simply a speeded-up version of the conventional shear. Speed may be increased 75% or even 100% for special purposes. Usually, this kind of equipment is used for cutoff work or long-run production, working from coil stock. The actual speed of the shear, in most manual squaring work, is not a significant factor because the shear is usually capable of working much faster than the operator can feed it.

The second type is high-speed equipment by almost any definition. The Famco PC line, an excellent example, is capable of shearing 12-gage mild steel, 72 in. wide, at 300 spm. The major application for this kind of equipment is in high-speed automatic cutoff lines. Since it completes a stroke in 0.20 sec, there is very little humping of the stock.

The cutting action seems to be improved at high speeds. This phenomenon has been observed many times in shearing and in high-speed press-

working; a number of explanations have been advanced. The most popular theory is that greater penetration takes place before fracture occurs.
In pressworking, it is known that a dull blanking punch that produces
severe burrs at 125 spm will make perfectly acceptable blanks if the speed
is increased to the 400–500 spm range.

High-speed shears have been very successful in blanking silicon steels
and a number of synthetic materials. This success seems to be a function
of the speed of the knife.

Many special shears have been developed for shearing masonite, clad
and exotic metals, wire mesh and unusual shapes, stone, rubber and even
burlap. For most metals, such as corrugated roofing, shaped siding, re-
inforcing mesh, etc., the shears are special only to the extent of relatively
inexpensive modifications to standard shears. Given the problem of shearing unusual metals or shapes, the fabricator should check with builders.
In most cases, the necessary engineering will probably have been done.

Chapter 13

HOLDDOWNS, GAGES
AND OPTIONS

Firm holddowns and fast and accurate gaging methods are essential to good shear performance. This chapter discusses them as well as a range of useful options that can greatly increase the productivity of the shear and the quality of the work produced in terms of accuracy and squareness.

Good firm holddowns and precise gaging devices are essential to accurate shearing. Without them the accuracy built into the shear itself is wasted.

The purpose of the holddowns is to lock the material being sheared solidly against the bed. This requires a great deal of force. Remember Murphy's Law? If a part can fail, it will fail. A similar axiom applies to shearing. If the material can shift, it will shift. As the upper knife makes contact, there is a strong lateral thrust added to the lever effect of the ram. The holddown must be strong enough to resist both of these forces.

The shear buyer has few options with regard to holddowns. The various builders have usually settled on one type of holddown, which is standard on their line. They may use springs, mechanical action by means of cams or toggles, straight hydraulics, air cylinders and combinations of methods to provide holding force. These are designed with ample reserve margins. Holddowns are seldom a source of any trouble.

The user should expect to be able to cut polished stock without excessive marring of the sheet by the holddown fingers. He should be able to cut pieces of varying thicknesses in one stroke of the shear. He should be able to cut short pieces even if engaged by only one plunger.

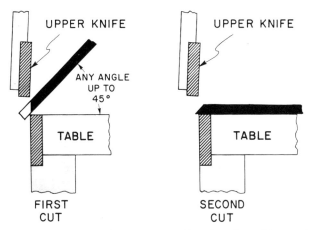

Bevel shearing on Cincinnati shear equipped with adjustable angle supports. Plate can also be sheared to a point by reversing. This is a true test of shear performance, holddown power and lack of deflection.

Available holddown options include rubber or plastic feet for the holddown plungers, individually adjustable plungers and special combination holddowns that can be used in shearing delicate material such as stainless-steel honeycomb. For most jobs, the stock holddown furnished by the builder will be quite adequate.

Gaging

Probably 90% of the inaccuracy problems encountered in shearing result from bad gaging habits. If care is used in seating the stock against the gage, if gage settings are frequently checked, and if the correct gaging procedure is used, most of these problems will solve themselves.

Of the three gaging methods, back gaging is used most often. The back gage is nothing more than a stop against which the stock is located. Usually it is operated from the front of the shear by means of a crank. Most back gages are calibrated to read $\frac{1}{64}$-in. or $\frac{1}{128}$-in. increments. The human eye cannot readily resolve divisions finer than this.

For production work, a power back gage is strongly recommended. It will usually pay for itself in a very brief time in reduced waste motion. Power back gages are furnished as standard equipment by some builders on their medium and large shears. For shearing heavy stock, a hinged back angle is recommended. This can be swung up out of the way to allow work to be passed through the shear. Thin sheet is usually flexible enough to be passed under the gage so that a hinged angle is not required.

Front gaging is accomplished either by locating the work against stops set into the table or into support arms at the front of the shear or by

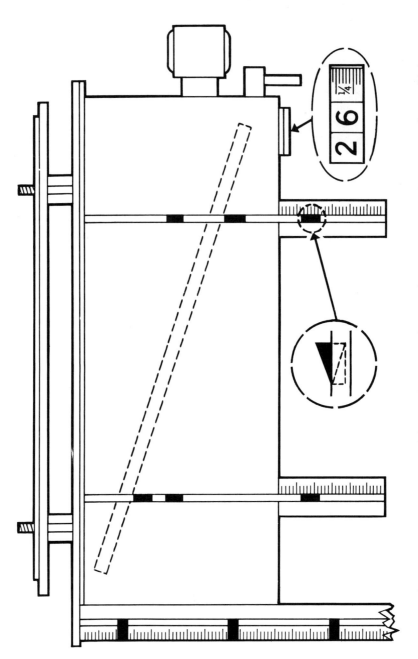

Gaging layout shows the many possibilities with proper selection of options. Text in chapter 15 refers to this layout.

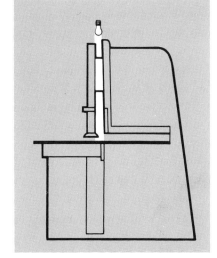

Power back-gaging, working from a direct-reading dial at front of shear, saves operator wasted steps.

Light beam, projected to throw a sharp shadow on scribed line, eliminates need for end sighting.

locating the side of the stock against a squaring extension. Table, support arms and squaring extensions are provided with adjustable, highly accurate scales. Neither method is quite as fast or quite as accurate as back gaging. Both are useful on some jobs.

Front gaging, using stops on the table or the support arms, for example, should be used when shearing small quantities of thin stock that might

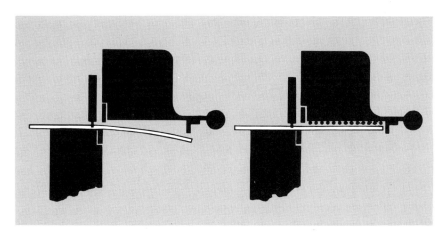

Magnetic sheet supports prevent sagging when working with thin material so that more economical back-gaging method can be used.

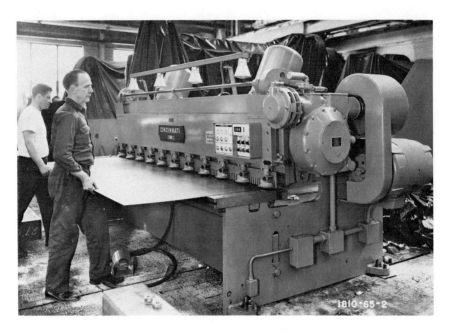

High-speed drive on this shear is optional on many of its maker's shears. It's capable of 100 spm on shears with capacities through ⅜-in. mild steel. With automatic knife cycling, controlled by probes, it makes high order of automation possible.

have a tendency to sag between the knives and the back gage. It can be used to shear lengths beyond the capacity of the back gage.

Disappearing front stops mounted in slots in the bed or support arms are highly useful. These allow the stock to pass forward. When the stock clears the stops, they are raised, usually by spring action, and the stock is located against their forward edges. Work handling and the possibility of marking the stock are reduced. A front gage bar can also be used to cut angles with excellent accuracy.

Power-driven back gaging, disappearing front stops, calibrated squaring extensions in lengths up to 12 ft and extra-long front support arms are usually available either as standard equipment or as options. They are relatively inexpensive and will almost invariably pay for themselves regardless of the type of operation for which the basic shear is bought. They not only speed up most operations—they also greatly increase the versatility of the equipment. A good shear operator can perform prodigies with them. An average shear operator can hardly operate without them.

Automatic Cycling

Automatic cycling is now available as an option on some shears. Coupled with high-speed shear drives and power feeding, this is a very effective mass production "package." The Cincinnati, Incorporated version is an excellent example. Available on all the company's mechanical shears through $\frac{3}{8}$-in. capacity, the control is actuated by three electronic probes recessed in the back-gage angle. Any two of the probes can be used to actuate the shear, depending on the width of the work being cut. Selection of the appropriate probes is made from the front of the shear. Since two probes must be touched by the sheet simultaneously, the cut must be parallel.

Several other knife cycling methods have been developed. They differ mainly in the method of actuation.

Shear Feeding

Other than the high-speed models used in cut to length lines and fed from coil stock, few shears are provided with feeds. The general run of job-shop work with extremely short runs does not lend itself to other than manual feeding. For the occasional long-run job or for warehouse operation, feeds can be devised. Conventional press feeds of the hitch or gripper type have been adapted successfully to feeding sheet material. Accuracy is usually not very good since the length of the off cut is determined by the feed and the stock is not located positively against stops. The development of automatic cycling, described earlier, will create much greater demand for shear feeds, primarily because of the excellent economics of coil vs sheet stock and the more consistent coil quality now available.

A limited group of special shear feeds is available. Usually they operate by means of a magnetic roller. The sheet to be sheared is trimmed on three sides, using the side and back gages. The untrimmed edge is then fed through the shear until it clears the knife. The magnetic roller feeds the sheet forward against the front gages. The cycle is automatic until the sheet is completely sheared.

This approach to automatic feed (American Actuator) uses magnetic rollers. Note position of the stock with relation to rear of shear frame.

Magnetic feeds of this type can be interlocked with the shear circuitry so that the clutch is engaged and the ram cycled when the leading edge of the sheet comes in contact with the front gages. There is no limit to the width of stock that can be sheared in this manner other than the width of the shear itself but the method cannot be used where the offcut is less than 3 in. By using an added pinch roll, magnetic drives of this type can shear nonferrous material.

Other Options

The more important additional optional equipment for use with shears includes remote control, rear sheet support, light-beam shearing gage and, for heavy work, ball transfers in the bed.

The remote control option should be picked up when large sheets will frequently be handled. In this case it is often difficult for the operator to use the customary treadle.

Magnetic sheet supports should be specified if a considerable volume of steel 14-gage and under is to be sheared. The magnetic supports hold the stock in a level plane, eliminating the sag that can be a nuisance in light material. It is possible, when sag is a problem, to use the front gaging method. For relatively small quantities this is practical. As volume increases, economics favor the faster back-gaging method and the purchase of magnetic sheet-supports is justified.

The light-beam shearing gage is a useful accessory for bulky work and for short-run shearing to scribed lines. It consists of a light source, care-

Operator merely removes sheared blanks. Cycling is automatic. Stock is almost completely sheared. Reversing feature drops off cut onto pallet.

fully focussed so that it corresponds to actual knife position. The operator sets the scribed line to the shadow. It is unnecessary for him to sight from the side of the shear.

A variation is a mirror light gage. This includes a light source and a mirror arrangement that shows the operator the scribed line and the knife face at eye level. This also eliminates end-sighting and is useful in shearing angular or irregular shapes to scribed lines.

Other useful options include bevel shearing attachments, stroke counters, power operated front gages, power operated squaring arm stops, preselect gaging, special material delivery chutes, vernier pin gaging, rear view mirrors, and pendant controls. Some of these options are discussed in detail in Chapter 16.

It should be apparent that the modern shear is not merely a means of cutting stock roughly to size. It is a precision tool, capable of excellent production rates and a high degree of accuracy. With the proper accessories, it can be tailored exactly to a specific job or a range of work.

Chapter 14

SELECTING AND USING SHEAR KNIVES

The selection of knives is complicated by so many factors—type of material, thickness, temper, etc.—that most shear users should rely on the shear-knife makers for guidance. The many considerations are discussed, and some practical suggestions that will increase the life of the knife are given.

The finest shear that money can buy will not give economical and accurate production unless the knife is of equal quality. Bargain knives are never really bargains. Nor is do-it-yourself sharpening unless adequate equipment and skilled personnel are available. When the ram of the shear descends, tremendous forces are exerted on the cutting edge of the blade. The slightest defect in the metallurgical structure of the knife or the smallest shortcoming in the heat treatment or grinding processes can cause it to fail. At best they will result in excessive knife wear, premature failure and unsatisfactory shearing.

Fortunately the problem of knife selection and use is not as great as it once was. Not too long ago, the knife not only had to be sharpened periodically—it also had to be rehardened. The user had to rely on the skill of the heat treater—usually this job was done outside because of the rather large facilities required—as well as on that of the grinder. Then (as too often now) the heat treater usually operated by instinct, rather than by instrument. He relied on his eye to judge the correct quenching temperature. With all due respect for the skill of the old timers, he could not really judge temperature to within the few degrees that are now considered critical in heat treating.

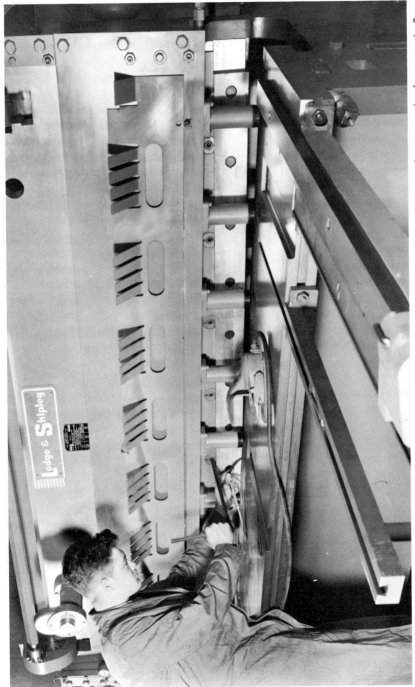

Use of feeler gages insures correct alignment of Wapakoneta knife in modern shear. Can you detect good safety practice?

Modern deep-hardening steels have eliminated the rehardening problem. If a few important rules are observed, the shear user can expect longer and more profitable service from the knives he buys than was ever before possible.

Selecting the Knife

There are so many grades and hardnesses available today, most of them listed under proprietary names, that the business of selecting a knife would seem to be confusing. It needn't be. The fact is that so many factors enter into the choice (initial price should not be one of them) that very few shear users are competent to select a knife for a specific application. This should best be left to the maker or supplier of the knife unless the services of a metallurgist are available. Even then, the knife maker will probably be far more competent to make recommendations.

He will want to know the maximum thickness, condition and type of material being sheared. He will want to know whether shearing is done continuously or occasionally. He will then be able to judge whether an analysis that will provide high shock resistance or one with better wear resistance will be most economical. If a wide range of thicknesses is to be sheared with a single set of knives, he may recommend a compromise. Some of the wear resistance needed for cutting light gages may have to be sacrificed for the shock resistance needed in cutting heavy plate.

How Knives Are Rated

Knives are rated according to their shearing capacity in mild steel or its equivalent in tensile strength. For example, knives engineered to shear mild steel $\frac{1}{4}$ in. thick could safely be used to shear 310 stainless steel 0.200 in. thick or type 3S-H14 aluminum 0.900 in. thick. However, users should not make the mistake of assuming that a mild steel plate $\frac{1}{2}$ in. thick × 5 ft wide can be sheared with knives designed to shear stock $\frac{1}{4}$ in. thick × 10 ft wide simply because the final sheared area is 30 sq in. in both cases. If the shear being used has a rake of $\frac{5}{16}$ in. per ft, the actual area under shear at any one time is $\frac{1}{4}$-in.—thick material would be approximately 1.2 sq in. as opposed to 4.8 sq in. in the $\frac{1}{2}$-in.-thick plate. This is a common error and usually results in early destruction of the knife. If the shear has an adjustable rake feature, of course, it is possible, within limits, to compensate for various thicknesses.

Maintaining Shear Knives

It is essential that shear knives be kept sharp. It is strongly recommened that an extra set of knives be kept available for each shear. The extra set of knives makes it possible to pull the used knives before they become excessively dulled. This is best done as part of a controlled re-

Wooden blocks will protect Wapak knife if it's accidentally dropped. This is good procedure and should be followed.

sharpening program. If extra knives are not available, the temptation to "keep running" on the installed set is very great. No foreman wants to run up downtime when he can avoid it.

There is an economical point at which knives should be resharpened. If the knives are used beyond this point, they become dull and the cutting edges become fatigued. As fatigue progresses, dulling becomes more rapid, in a sort of vicious cycle. Eventually the edges spall. They may break or chip. In any event, excessive metal must be ground off to bring the blade back into good condition. The life of the blade may be immeasurably shortened.

In addition to chipped edges and breakage, dullness can cause heavy burrs on the sheared edge of the stock. Dull knives cause excessive wear on the ram guides and may also seriously overload the shear. If the shear is operating at near capacity, dull knives may cause major damage to the shear structure itself.

Many users who would not dream of using a dull milling cutter, because they realize that this is inefficiency at its worst, will continue to use dull shear knives. Perhaps this is because the size and configuration of a shear knife make it easy to forget that it is a precision cutting tool—just as much as a milling cutter or a set of punches and dies. It deserves as much care as any other precision tool.

Grinding the Shear Knife

Grinding a shear knife is a demanding process. It should not be attempted unless a precision surface grinder of adequate size is available. Even if such equipment is available, the job requires personnel skilled in grinding large sections of hardened tool steel to precision tolerances. If the equipment or the personnel are not available, in-plant grinding of shear knives should not be attempted. The quality of the cut made by the knife will be poor or unacceptable if the knife is not ground to exceptional parallelism throughout its length. This job is far better left to knife makers who offer grinding service—most do—or to grinding specialists. This is another excellent reason for stocking an extra set of knives—it won't be necessary to take a shear off the line while the knives are being ground.

If the knives are to be ground in-plant, the grinder should be in first-class condition. Any inadequacies in the grinder will surely be reproduced in the knife. The type of abrasive to be used varies with the grade and hardness of the knife and with the wheel and table speeds of the grinder. Generally, aluminum-oxide abrasives in a vitrified bonded wheel will give the best results. The grain size and wheel hardness should range from 46 grain, H hardness for plate shears to 60 grain, G hardness for knives designed for cutting light-gage sheet. The structure should be no denser than No. 8. The wheel should not be allowed to load up. It should be

Proper grinding of knives takes massive, accurate equipment like this Mattison grinder. Ohio Knife Co. photo.

dressed frequently to make sure that it remains free cutting. The knife being ground should be constantly flooded with coolant at the point of wheel contact.

Light feeds should be used—not more than 0.001 in. in most cases. The feed should never be great enough to cause any heat discoloration of the knife. If this happens, the damage is permanent, even if the discolored portion is ground away. At best, there will be a loss of hardening in the discolored area. At worst, there will be heat checking or buildup of severe internal stresses. Either will cause chipping of the cutting edge and may cause cracks through the entire section.

Before establishing a knife-grinding program in-plant, users should determine that they can match the accuracy of the grinding performed by the knife maker. The Wapakoneta Machine Co., for example, grinds its blades parallel within 0.005 in. in width from end to end and parallel within 0.003 in. in thickness from end to end. There is no variation greater than 0.001 in. within any 12-in. length. These figures are typical of the high standards of the knife producing industry. Obviously, they will be difficult for the average using plant to match. Further, many knife makers maintain files showing the metallurgical and physical characteristics of each knife produced. With this information, they are in an excellent position to establish ideal grinding conditions when their product is returned for resharpening.

Installing Shear Knives

The same care that is taken in mounting any precision tool should be taken in installing shear knives. The knife seats must be clean and free from burrs. Otherwise the knife cannot possibly be seated rigidly and firmly. For the same reason, the knives themselves must be wiped clean just before installation. The shear knife is relatively fragile when it is unconfined. It becomes enormously strong only when it is properly confined and bolted to its retaining members in the shear. A burr, a chip or a particle of dirt can literally cause the ram to act as an enormous shearing load.

The shear knife should be handled with care. A knife will chip easily, if dropped. A passing fork lift truck can cause severe damage with just a glancing blow. The knife should be kept in its box, fully supported, except when a change is being made. The use of a sling, designed to fit the bolt holes and to distribute the weight of the knife over a wide area, is recommended.

Although bolt holes for mounting knives are standardized, most shear manufacturers recommend specific routines for knife change. These should be followed. If a copy of the appropriate manual is not on hand, it should be obtained. Common sense, of course, is a fairly good substitute. The

Vertical atmosphere furnace used to harden Tool Steel Gear and Pinion (TSP) shear knives can accept 22-ft knives. Heat treating in vertical furnace tends to produce optimum straightness in knife.

upper knife edge should be supported on wooden blocks while being changed. The lower knife should be shimmed to the exact level of the table after regrinding. Most modern shears provide adjustment of the crosshead height to eliminate shimming of the upper knife.

The shear should be leveled on installation. The level should be checked frequently, remembering that the very nature of the break-through shock is apt to tilt the shear or to wear the foundation unevenly. Many a shear engineer has travelled a country mile in a hurry because "that piece of junk you sold me won't cut a straight line." The experienced man will check the level of the shear as his first move. More often than not, no further investigation is needed. Check the level every time a knife is changed as a matter of routine.

Knives not in use should be protected with a heavy coat of rust preventive oil. They should be stored, in their boxes, in a dry area, free from sudden temperature variants that could cause condensation to form in the box. Shop people should be discouraged from using the boxes as convenient storage racks for oil drums, spare gears, and all the other objects that seem to accumulate in any shop on any flat, convenient surface.

Knife Clearance

As mentioned in an earlier chapter, shear builders do not agree on the most efficient knife clearances for various thicknesses and kinds of material. As a general rule, unless a set of arbitrary clearances has been developed, the clearance should be set at 7% of the thickness of the material being sheared. In making cuts at or near the capacity of the shear, the center of the upper blades should be bowed 0.002 in. nearer the lower blade than at the ends. The purpose is to overcome the deflection of the ram under high loading conditions. Some shear makers recommend or provide alternate methods of overcoming deflection. Consult the manual on this subject.

Rule-of-thumb clearances for shearing some common metals are:

Metal Thickness (In.)	Mild Steel	Clearance (In.) Brass, Aluminum Stainless, Copper, Silicon Steel
$0.005-\frac{1}{32}$	0.002	0.002
$\frac{1}{16}-\frac{3}{16}$	0.005	0.002
$\frac{1}{4}-\frac{1}{2}$	0.017	0.008
$\frac{5}{7}-1\frac{1}{4}$	0.043	————

Ordering Knives

In ordering knives, the knife maker should be provided with the following information:

1. The maximum thickness of the material to be sheared, regardless of the amount of material to be cut.

2. The type of material to be sheared: mild steel, alloy steel, stainless steel, or nonferrous metal such as aluminum, brass, copper, etc.

3. The temperature of the material. Is material sheared at room temperature or hot? If hot, state the maximum temperature and if water is used.

4. Condition of material. Is it hot-rolled, pickled, etc?

5. The type of operation. Is it intermittent or steady? Does material thickness vary widely?

6. The make and model number of the shear.

7. A print or a simple sketch showing all dimensions affecting the size of the knife and provisions for mounting.

If the procedures set down in this chapter are followed carefully and conscientiously, the result can only be longer and more efficient knife life.

THE THEORY AND PRACTICE OF SHEARING

This concluding chapter discusses shearing theory from a practical basis. It also establishes rules for good shearing practice and reviews some of the factors that should be considered in selecting equipment and accessories.

The shearing process is extremely complex—so much so that entire books have been written about it. Such factors as crystalline structure, slippage planes, brittle fracture and anisotropy are involved. These are of interest primarily to the metallurgist. At the same time it isn't enough for the shear user to say, "The knife goes about a third of the way through and then the piece breaks off," even though this is essentially true and can be confirmed by examining the edge of the off cut.

It is known that both tensile and compressive stresses are involved. As the knife contacts the metal being sheared, the top surface is momentarily under tension, while the bottom, supported by the lower knife, is under compression. As the elastic limit of the metal is exceeded, it is stressed in shear until its ultimate strength is exceeded and the piece fractures, breaking completely away from the parent metal. If knives are sharp and the clearance is correct, the sheared edge will be clean, with little burr and will be very close to the perpendicular.

A series of super-speed motion pictures, made under the sponsorship of Niagara Machine & Tool Works, has successfully "fixed" the action that takes place. It demonstrates that the final fracture is quite violent in nature, with the off cut almost literally exploding away from the stock. The film also demonstrates the need for good firm holddowns. In its recovery

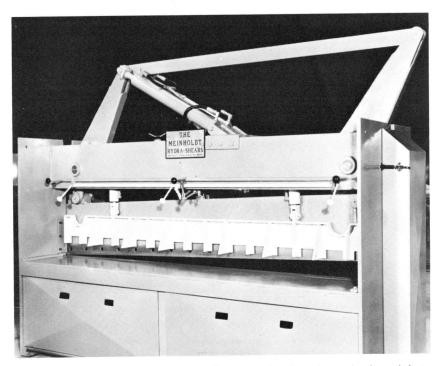

Unusual design of Meinholdt Hydra-Shear permits changing rake from left to right to right to left by changing position of the hydraulic cylinders.

from the compression strain, the narrow section between holddowns and knives exerts a strong upward pressure.

The internal stresses that are present during shearing explain the three most common bugaboos in shearing. These are bow, twisting and camber. Twist is perhaps the most common condition. Twist is the tendency of the off cut to curl into a spiral or corkscrew. Once twist is present in the off cut, it is difficult to remove. Twist is definitely a function of internal stresses but it is also a function of the rake of the knife. The less rake, the less the off cut will twist.

Rake and knife clearance were discussed briefly in a previous chapter in terms of equipment selection. They can be more fully explained by analogy. An almost exact parallel can be drawn between a shear, a pair of scissors, and a pair of tin snips. The actions are identical.

Try to cut a piece of heavy cardboard with the scissors. If it can be done, it takes a great deal of leverage. Cut the same cardboard with a pair of tin snips. They will make the cut easily. This is because the angle (rake) between the blades of the snips is large so that the shearing action takes place over a limited area. The scissors have a much flatter angle. Cut a

Arranged for cutting heavy plate with minimum effort on the part of the operators, this shear typifies modern equipment. It's equipped with virtually every

option and safety device that can add to its productivity. Ball transfers, for example, eliminate bull work, let operator concentrate on accurate gaging.

Most shears have ample safety margins if properly used. Turning a hand lever 180 deg on this Niagara model positively prevents the clutch from engaging.

piece of paper with the scissors and it will lie flat. The piece of cardboard cut with the snips will have a pronounced twist. Similarly, the higher the rake angle in a shear, the greater the tendency of the stock to twist.

It is also true that the narrower the off cut, the more it will twist. Heavy stock will twist more than thin stock and soft stock will twist more than hard material.

The same analogy can be drawn for knife clearance. Cut a piece of paper with scissors. Now loosen the pivot bolt between the blades. If you try to cut with them now, the paper will probably jam between the blades. This condition is identical with too great knife clearance and will have precisely the same effect. Oddly enough, the direction of shearing seems to have little effect on twist. If the knives are sharp and properly set, twist-

Demonstrating again the versatility of modern equipment, this 30-ft hydraulic shear is used for short-run work ranging from 11-ga to 1/2-in. plate.

ing can be kept to a minimun. If it is excessive under these conditions, the trouble probably stems from poor quality stock.

Camber is the tendency of a sheared strip to take on an arc while lying flat. The rake of the shear has little effect on camber. Camber is usually caused by internal stresses, inferior material or improper knife adjustment. Inferior material is the greatest offender. If camber is severe after levelling and knife adjustment have been checked, try to shear a piece of stretcher-levelled stock. Almost invariably the camber will disappear and the cut will be perfect.

As with twisting, camber is most severe on narrow strips. When shearing material into strips, prime-quality material should be used.

Bow is the tendency of a piece of sheared material to hump in the center.

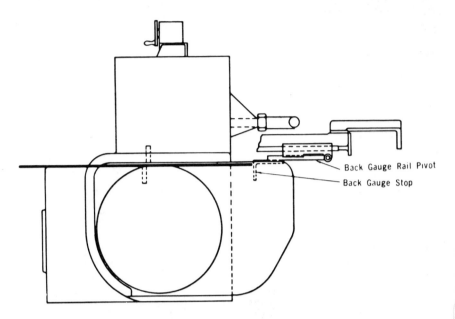

Hinged back-gage, available on virtually all shears as on this Wysong model, can be swung up to permit shearing beyond the range of the back gage.

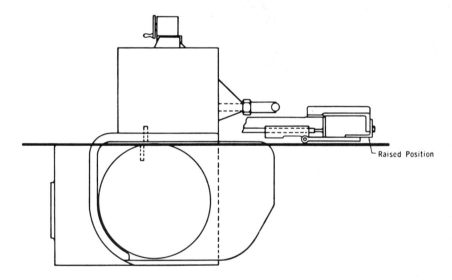

Back-gage in the raised position. Power back-gaging from the front of the shear is usually offered as an option. In most cases, it should be specified.

In the well-organized shop, the shear department foreman usually lays out the work. He determines which gaging method should be used and what the cutting procedure will be. On fairly large orders, this is by no means a simple problem. It's so complex that it actually has a name—the two-dimensional cutting stock problem. It's defined as "finding the least wasteful and most economical way of cutting large sheets of stock into smaller rectangles to meet incoming orders." Given a large number of orders, there are literally millions of possible combinations. Very recently, IBM Corp. developed a program for the IBM 7094 computer aimed solely at solving this problem.

The average shear user will never require a computer to lay out shearing jobs, but the problem is still a difficult one. There are more than 240 ways of shearing a 3 × 6-ft sheet into 3 × 6-in. blanks, for example. More than half of them are wrong in that they waste stock. More than three quarters of them are wrong because they take extra operations. Here are two very common shearing applications with recommended methods. Note that many other methods are possible.

Shearing Small Blanks From Rectangular Sheets

Assuming a 3 × 6-ft sheet is to be sheared into small rectangular blanks, the best procedure in this case will usually be to trim one long edge of the sheet by gaging the end against the squaring arm. Judgment now enters the picture. Whichever dimension of the small rectangles divides most evenly into the width of the sheet should be set on the back gage. The sheet should then be sheared into strips, using the back gage. The sheared strips are now trimmed, using the squaring arm. They are fed through and gaged against the back gage. (The value or power operated back gages should be readily apparent on this kind of work.) The last strip will have to be turned around and gaged against the front stops. A skilled operator, seating the stock firmly against the gages, can work to within 0.005 in. and better with this method.

Shearing Large Blanks From Thin Stock

This is a common situation and it creates a problem because thin sheet will sag if it is not supported. This makes the use of back gages impractical. Usually the best procedure will be to trim one long edge of the sheet by gaging aginst disappearing two-position stops in the support arms. Now one short end can be trimmed square by locating the trimmed long edge against the squaring arm and gaging against a preset stop therein. The remaining long edge is trimmed against the disappearing stops and the fourth trim cut is made by gaging against the stop in the squaring arm.

We now have a square sheet. It can be cut into blanks of the required width by gaging against swinging stops in the squaring arm. This job re-

Again, this is almost always due to deficiencies in the material. If
is considerable variation in thickness across the sheet, bowing is a
certain to result and little corrective action can be made at the shear.

Tonnage Ratings of Shears

Shears are rated on the basis of their ability to shear mild steel of a
thickness and length, much in the manner that shear knives are rated
manufacturer's capacity rating should not, under any circumstance
exceeded. A shear rated at $\frac{3}{8}$-in. × 12 ft should not be used to cut
stock even if the length is smaller than 12 ft.

The load on the shear is a travelling load and remains constant, i
less of the length of the cut. The load on a shear, in cutting mild st
creases almost in proportion to the square of the metal thickness
overloads on the order of 200 to 300% can be caused by fractional in
in metal thickness.

Tonnage requirements are a function of the shear strength of the
to be sheared. A shear capable of shearing 16-gage mild steel with
strength of 50,000 psi cannot be used to shear 16-gage stainless
shear strength of 75,000 psi. At the same time, it can be used to s
gage soft aluminum with a shear strength of only 9,000 psi or
aluminum alloy with a shear strength of 20,000 psi.

Although there is a relationship between shear strength and
requirements, there is considerable difference of opinion as its exact
It is not a straight line relationship so that a conversion factor is
Many shear users have developed their own arbitrary factors.
shear builders have developed tables showing relative tonnage
ments of mild steel and a number of commonly used metals. The
been worked out after lengthy research and are not available for
publication. If you intend to shear metals other than mild steel, t
builder should be asked to set down maximum thickness limits
metals you wish to shear.

Developing Sound Shearing Practices

Work for the shear should be carefully planned. The object
maximum utilization of the sheet. It is a treat to visit a well-o
shop. Assume that such a shop is shearing a number of square
blanks from 0.060-in. mild steel. The order will have been made o
production scheduler. If he is doing his job well, he knows the e
and quantity of tail ends in the stock room. Where the job requi
blanks, he will use up the tail ends and leftovers before ordering f
from stock. If he has a considerable quantity of small pieces of
gage number, over or under, he may call the customer and ask if
tution is permissable.

quired more, slower and less accurate operations than back gaging. If the quantity of thin material being processed is large, magnetic sheet supports should be specified. With them, even thin ferrous material can be cut more rapidly and more accurately, as mentioned in Chapter 13.

Maintenance and Safety

No hard and fast rules can be set down for shear maintenance. The common sense practices that apply to any precision machine tool should be followed. All shear manufacturers supply maintenance manuals with their product. Many of them are quite complete and contain a great deal of useful information. A maintenance schedule in keeping with the manual should be set up and adhered to.

Good practice requires that the shear be located centrally to reduce materials handling. This is a much neglected practice. In many shops, it sometimes appears that shears are placed wherever there happens to be an open space. If a shear is serving a battery of presses, the traffic in and out of the shear area will be very heavy. Imaginative engineering can do wonders here. By relocating the shear, it may be possible to set up simple conveyor systems that will substantially reduce materials-handling costs.

Properly used, a shear is a safe tool. Statistics indicate that there are relatively fewer accidents involving shears than power presses or press brakes. Most of the accidents that do occur are caused by carelessness and poor housekeeping, in that order.

Carelesness cannot be eradicated. It can be reduced by an adequate safety program. Poor housekeeping is inexcusable. The shear that is installed off in a corner is apt to be a dirty shear—perhaps because the supervisor doesn't pass by it every hour or so. Trim is allowed to accumulate. Dirt also accumulates, sometimes in sufficient quantity to effect the performance of the valves and electrical controls that are the operator's chief safeguard. The dirty shear, badly maintained, is an inefficient mechanism. The well maintained shear, properly adjusted to its job and used as any precision tool should be used, is one of the most productive and accurate machines in the plant.

Chapter 16

NEW DEVELOPMENTS IN PRESS BRAKES AND SHEARS

Since the first edition of the Press Brake and Shear Handbook was published, a number of significant developments have been made both in the equipment itself and in the technology of bending and shearing. Some of them are of obvious importance; others, although seemingly of minor degree, can contribute heavily to better equipment utilization, reduced downtime, and a more accurate product.

Shearing by Numerical Control

Numerical control (NC) is unquestionably the most important development in shearing. Available from several shear manufacturers, it has a number of advantages. They have been nicely summarized by Cincinnati, Incorporated, which was the first of the shear builders to offer NC and which at this writing is marketing its second generation of NC shears.

1. NC makes possible direct management control of shearing operations. Random shearing requirements can be organized by computer and sheared at the least cost per blank. The necessary computations to do this by manual calculation would not be economically feasible.

2. NC virtually eliminates operator planning and setup, which can often take eight to ten times as long as the actual shearing operation.

3. Computerized NC makes maximum use of material and reduces scrap to the lowest possible amount.

4. NC makes it possible for management to standardize the size of input sheets, which reduces inventory and lowers material purchasing costs.

5. NC shears can outproduce conventionally controlled shears on the

order of 2 to 1 and more. This makes plant floor space savings possible and, obviously, reduces labor costs.

6. With its remarkable accuracy and, particularly, repeatability, NC shearing is extremely compatible with other processes such as slitting, cutting to length, hole punching, and so on.

The NC shear should not be measured against a manual shear on a one-for-one basis. In the first place, it requires a considerably higher capital investment. In the second place, if the potential of the equipment is to be realized, the NC shear should be regarded as a system. The criteria for evaluating NC have to do with the type of operation. Generally, an auto-mated NC shear should be considered where the shearing department satisfies one or more of the following conditions. It:

a. Shears from 1 to 100 pieces of many sizes.
b. Shears from 25 to 250 sizes in several gauges.
c. Shears 10 to 20 gauge material in any quantities.
d. Has three or more standard shears.
e. Has a frequently changing shearing schedule.
f. Has a high small-lot scrap rate.

A brief description of a typical NC shearing system may help the reader to visualize how it works and why this approach is significant. Niagara's Niagaramatic system was originally developed to complement an NC turret punch. NC turret punching machines are extremely fast and accurate and can produce enormously complex hole patterns with superb repeatability. The Niagaramatic was designed to either prepare blanks for NC turret punch presses or to cut the punched sheets produced thereon into segments. Speed and accuracy are essential elements of the system—for instance, no cumulative error can be tolerated.

The shear itself is an underdrive model with remote tripping and auto-matic lubrication. It has a front-mounted feed table, an extension squaring arm and a rigid box-section positioning carriage. The carriage has three air-operated sheet clamps with quick disconnects for rapid positioning to accommodate various sheet widths. The carriage rides on two round horizontal ways fitted with precision linear ball bearing bushings and is located by a numerical control working through an electronic rotary re-solver. The carriage is actuated by a DC servo motor through a timing belt drive to a hardened, precision lead screw with preloaded recirculating ball bearings.

The control itself is of the single-axis type and works from a punched tape through a tape reader in the operator's control console. The control has absolute dimensioning to eliminate cumulative error and is completely transistorized. Programming is quite simple and does not require a computer although several Niagaramatic installations have been tied in to computer centers. Even on very complex jobs, the tape itself seldom exceeds 18 inches in length.

With a programmed tape in the tape reader, the operator places the sheet to be cut on a feed table in front of the carriage. He starts the cutting cycle by simply pushing a button on the control console or on a remote station. (For full automation, the table can be fed from a sheet loader.) The carriage clamps then grip the material and the carriage traverses at high speed to the exact cutting position directed by the tape at which point the cut is automatically made. The sequence continues automatically until the predetermined number of cuts have been made. An integral stacker/conveyor/scrap separator separates the trim cuts and stacks the sheared sheets on pallets or stacker dollies. All operations following sheet loading are numerically controlled so that cumulative and operator errors are eliminated. Sheet handling problems are greatly reduced and a helper to stack sheets at the rear of the machine is not needed. This in itself is a safety advantage.

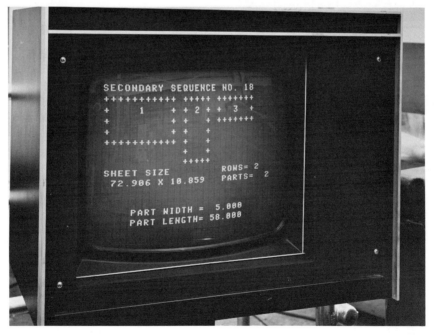

Advancement in NC shearing is this readout provided with a second generation NC control. Note that width and length readouts are to three decimal places. Accuracy is superb.

Pin Gaging—an Alternative to NC

Pin gaging, sometimes called pin shearing or vernier pin shearing, is a recent development that is nicely illustrated by Lodge & Shipley's Accra-Blank system. (Several similar systems are available from shear builders.) The Accra-Blank system was originally developed for use with NC hole-punching equipment and was designed in recognition of the fact that shearing to the previously sheared edge as a reference point had inherent disadvantages—chief among them being cumulative error.

A typical Accra-Blank machine has four arms, each with its own vernier scale and lock. The scales can be set at any point on the arm to an accuracy well within .001 inch. The reference point for gaging is a pair of preselected holes on the punched sheet. (On very small work, only one hole may be used in combination with an edge gage.) Since the holes to be used as reference points are generally round, tapered pins are used because they are self-centering.

Vernier pin gaging, shown here on a Cincinnati shear, is far more accurate than gaging from the sheared edge and is especially suitable for working with sheet punched on NC equipment.

Threaded holes are provided in the ends of the vernier arms. The locating pins are screwed into the threaded holes to the exact height desired—this is a function of the size of the hole in the punched sheet—and locked. The back gage has a function here; not as a locating device but merely to insure that the right holes are being used as reference points. A pair of pressure holddowns hold the sheet firmly against the pins during shearing.

Because most parts sheared from a punched sheet are further sheared in other dimensions, Lodge & Shipley has developed a compact conveyor that can be used to return sheared pieces to the front of the machine if desired.

Pin templates have been used for many years for work demanding particular accuracy. The cost of preparing templates and the difficulty of storing them has limited this method. Because the vernier pin gaging concept does not require templates, it is emerging as a fast, economical, and accurate method for producing precise blanks. As noted, accuracies well within .001 inch can be held.

Developments in Accessories

Developments in other, relatively minor, areas of shearing, as mentioned, can sharply increase the productivity of the basic equipment. Usually a minor capital investment is needed—if any. Some of these developments will be discussed and illustrated in the following pages.

Holddowns. Most shear builders have improved their holddown systems in recognition of the increasing trend toward the use of prefinished stock and also for safety reasons. One of the most interesting developments is the dual holddown, of which Lodge & Shipley's Hydro-Hold is an excellent example. Where impact with the material and noise are not a problem, the holddowns apply pressure at the beginning of the ram stroke. They are hydraulic and have ample pressure for heavy work. For work with finished or delicate stock, such as honeycomb, the fingers are brought down under shop air pressure until the stock is lightly held. Then only does the hydraulic system take over. Both the holddown pressure and the speed at which the fingers are brought down are controllable. Shear builders offer a number of variations on this theme.

Knife Change and Adjustment. Shear builders have recognized that knife maintenance is a major source of lost productivity. Some older designs were such that knife change was awkward and time-consuming; knives were not changed or removed for sharpening simply because it took so much time. It has also been recognized that certain mounting arrangements actually acted as stress raisers. A good example of shear design with knife maintenance properly considered is Niagara's standard mounting. Once the holddown is removed, the upper knife is fully exposed and can be

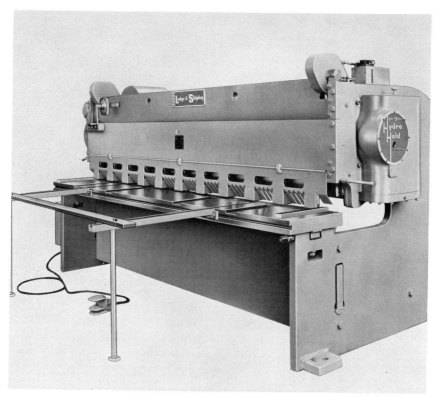

Advanced holddown provided with this machine uses soft touch of air to bring the holddown fingers into position, then applies heavy hydraulic thrust. It's excellent for soft, delicate work.

removed from the front of the machine so that no space is required for sliding the knife out at the side. The lower knife is secured by plow bolts and is readily accessible through openings in the top of the bed.

Some shear builders have recognized that many users of shears have neither the staff nor the equipment for proper knife maintenance. They offer a sort of turnkey service, which covers every phase of knife service from rotation to adjustment. For the small plant using one or two shears, this type of service is recommended.

The better shear designs now provide for simplified bed adjustment to compensate for the reduction in knife thickness through grinding. On some older designs this was an extremely complicated procedure and required a high degree of maintenance skills.

By making knives easy to get at—in this case it's only necessary to remove the holddown—better treatment and maintenance is encouraged. On this machine it is a matter of minutes.

Faster Setup. A number of approaches leading to faster setup have been taken. A typical example is the power-operated squaring arm stop offered by Cincinnati, Incorporated, on its shears. This device has a three push-button station control with which the operator sets the stop. It has two speeds forward and one reverse. The operator positions the stop against a graduated scale by means of a ½ horse power motor drive. Setup takes only a few seconds.

Improved Operator Information. Time was when the operator had to squint at a sparse selection of gages if he wished to know what was going on. There has been a universal trend among shear builders to provide high-visibility readouts at the front of the machine showing the operator

Builder of this shear has traditionally designed equipment for fast, efficient setup. Simple adjustment shown here moves bed laterally to compensate for reduced thickness of ground knives.

the status of every functioning component on the shear. More and more frequently they read in decimals rather than fractions, reflecting the increased accuracy of the equipment. Many shear builders, in addition to better information, now provide greater convenience in the form of pendant controls that give the operator all the information he needs plus complete operational control over all functions. The pendant controls can be moved within a reasonable range, making it possible for the operator to work from the most desirable position in relation to the operation being performed.

Feeds, Conveyors, and Stackers. The rising cost of labor, together with safety considerations, has led to the development of new families of sheet

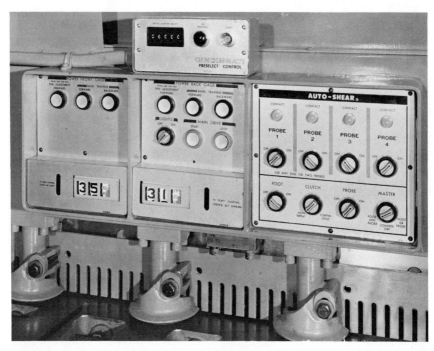

A major development in shears is the provision of all necessary information in highly visible readouts. They make it easier for the operator to operate his equipment to optimum rating.

and plate handling gear. A good example of an advanced feed is the AF-Series Shearfeed made by American Actuator Corporation. The Shearfeed connects to the shear through a supporting arm connected to the crosshead and moves up and down with the motion of the ram. The sheet or blank, regardless of size, is securely supported on the Shearfeed and the shear bed.

The operator sets the program so that the Shearfeed can provide any combination of cutting operations. While he is making primary or secondary cuts to the front gages, the remaining sheet is moved away from the shear blades to whatever distance is required and it will remain there until programmed to come to the operator or until he calls for it, which may be expedient when runs are very short or when single sheets are being cut. The operator cuts a blank or blanks from the full sheet, which is fed to either the front gages or to pins. He can then return the blanks to the feed for recutting into smaller sizes at different preset front gages in the same operation. This eliminates intermediate handling, which can be a substantial time waster. Blanks can also be squared and resquared using both

There are a number of advantages to pendant control as shown here. It's readily accessible to an operator at a remote station and is isolated from vibration and possible damage at operation point.

front and back gages. The blank is cut from the sheet at the front gages, squared, and recut into smaller sizes with the back gage.

All operations are performed by one operator at the front of the shear without a helper. He does not have to stop production to go to the rear of the shear to pick up pieces and the job of helper, often dangerous, is eliminated.

An equally interesting development by U.M.I. Corporation is a sheet stacker designed for use with high speed cut-to-length lines. It was built for

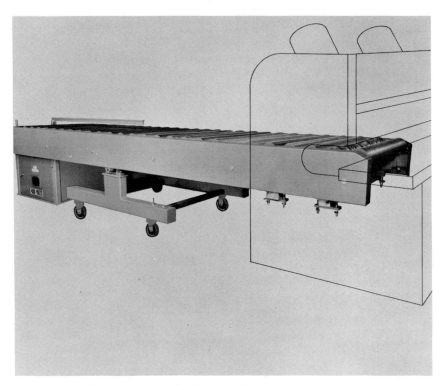

This is a very fast new shear feed made by American Accuator. It allows the operator to select any combination of cutting operations and can increase productivity to a phenomenal degree.

use with highly finished and even paper-covered stock in conjunction with a McKay shear line operating at 200 fpm. The sheet stacker consists of synchronizeb rubber-covered rollers that rotate in ball bearings supporting the material into the stacking station. As the stacker approaches the back-stop, the arms automatically retract, lowering the sheet to a conveyor. The conveyor is elevated and automatically lowers as the stack height increases. No pits are required and installation time is very short.

Several shear builders offer sheet handling equipment geared to their basic shears. The most useful approach is the systems approach, generally consisting of a conveyor, a separator, and a stacker. Cincinnati, Incor-porated, which offers a sheet conveying system, has made a study of machine utilization increases and labor reductions using the systems concept. Com-pared to manual handling of shear output using one operator, the utilization rate increased 55 percent while the labor reduction amounted to 59 percent.

These figures are cited to indicate the very great potential of sheet handling

This automatic sheet stacker by U.M.I., shown integrated into a high-speed shear line, handles sheet entering the stacker at up to 200 feet per minute and stacks it into a neat bundle.

systems. They are not suited to all shearing operations, but where they are compatible they merit the strongest consideration.

New Developments in Press Brakes

Major developments in press brakes in recent years have generally been aimed at reducing operator skills required to form accurate parts with a high degree of uniformity—in short, in the direction of more science and less art. In view of the great many variables involved in press-brake forming, a considerable degree of success has been achieved.

An excellent example is the Select-A-Form system developed by Pacific Press and Shear Company. The Select-A-Form is a programmable gaging system for press brakes that greatly simplifies and speeds up operations requiring the forming of multiple bends and angles in a single part. (Most

Plate conveyor used as adjunct to a heavy-duty shear can eliminate a great deal of bull work. It also makes working conditions more comfortable and safer for the helper.

earlier multiple gaging systems could not produce angle bends—these were formed in a separate operation.) The new system makes it possible for relatively inexperienced operators to set up and form simple or complex parts progressively in one continuous handling. Productivity increases are on the order of 60 percent as compared with conventional gaging. The ability to form complex parts in one handling greatly reduces the floor space and storage areas generally required for partially formed work.

Parts requiring as many as thirteen different bends can be programmed into the Select-A-Form system. Once set up, the gage will automatically position in a preset sequence as the ram cycles. Depending on the type of gaging system, as many as six different bend dimensions and four separate bend angles can be programmed in any desired sequence. The system is very accurate and the repeatability is excellent. Automatic multiple depth stops are provided. They have micrometer heads adjustable to within .001 inch. This makes it quite simple to home in on the desired bend shapes

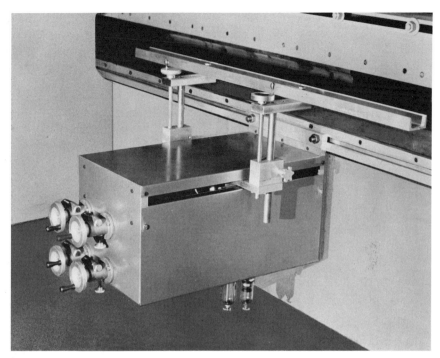

Important development in automatic press-brake gaging is this Pacific Select-A-Form system, which permits operator to preprogram as many as 13 different bends and up to 4 separate bend angles.

and to compensate for material variations so that much try-out and correction time is saved.

All stop positions can be independently adjusted from the rear of the unit and can be set precisely with digital indicators. Each setting handwheel and indicator enables the individual presetting of one stop position. The stop positions can be locked in place to prevent any deviation from the predetermined setting. For digital indicator accuracy, each stop position is preset to reference exactly with the distance between the front of the gage bar and the centerline of the press brake platen.

Angle bending capability is provided by multiple depth stops mounted on the right-hand end of the press brake. These make possible the preadjustment of bottom stroke settings to automatically form progressive bends having different angles. Depending on the system, from two to four multiple depth stops are available for progressive bends. Each depth stop can be individually preset and programmed to produce any desired bend angle. Large adjustments are made by repositioning the various individual assemblies. Fine adjustments are made by micrometer to an accuracy of .001 inch. The sequence of operation with regard to bend angle is programmed by selector switches on the control console. In operation, the depth stop settings are then automatically activated, one at a time, in the program sequence.

Once the job has been planned, the operator simply presets the back gage and depth stop settings and programs the control console by means of corresponding selector switches. He can then form a trial piece, check for accuracy, make any necessary corrections and proceed with the production run. Each piece can then be progressively formed in one handling because the back gages and depth stops will automatically program for each bend as the ram is cycled.

Developments in Controls

There is a strong trend toward the provision of more flexible controls by major builders of press brakes. A number of builders now offer fully auto-

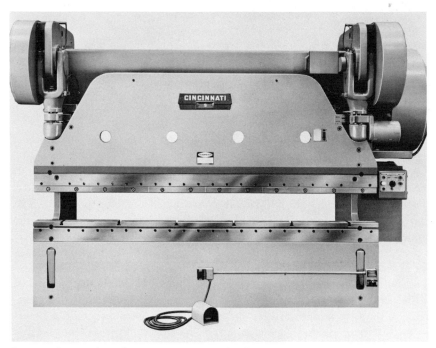

Improved control on this automatic press brake gives the operator more control options. Brake can operate as high speed equipment, low speed equipment, or automatically with dual speeds.

matic equipment that can be operated in any of several modes. Many now offer dual speed machines on which the percentage of stroke at high speed can be set through a rather broad range rather than at one or two fixed points. Major control developments, however, have been in the direction of high-visibility readouts, failsafe design, and the use of highly reliable components and circuitry. This, of course, reflects OSHA standards, which are discussed in detail in Chapter 17.

The final development in press-brake controls is NC. Several successful NC installations are in operation, although it would appear on the face of it that press-brake bending would not lend itself to tape. Cincinnati Shaper has had great success with NC and reports from the field indicate substantial savings and sharp reductions in manpower and in setup time. At one time it was generally believed that NC did not lend itself to shearing. Events proved that this was not the case—NC shearing has been highly successful. The much wider use of NC to actuate the gages for press-brake bending may not be far off.

Chapter 17

SAFETY STANDARDS FOR
PRESS BRAKES AND SHEARS

Late in 1973, the American National Standards Institute (ANSI) pub-
lished two new standards. They are National Safety Standard B11.3-1973,
Safety Requirements for the Construction, Care and Use of Power Press
Brakes, and National Safety Standard B11.4-1973, Safety Requirements
for the Construction, Care and Use of Shears. It can be expected that these
consensus standards will be accepted by the Occupational Safety and Health
Administration (OSHA) of the Department of Labor and published in the
Federal Register, probably in late 1974. Upon publication in the Federal
Register, they become official OSHA standards and have the force of law.

An Analysis of B11.4

National Standard B11.4 covers virtually all types of general purpose
shears. The only exceptions are certain types of rotary shears and flying
cutoff shears, which are not widely used. As such standards go, it is an
excellent piece of work in that it recognizes that there are limits to the
degree of protection that can be provided by the owner or the equipment
builder, and that a certain amount of common sense on the part of the
operator can be assumed. (Such an assumption was not made in the case
of a number of earlier standards.) The standard is also easier to read than
many consensus standards, which means that it will not be the subject of
endless interpretations and reinterpretations. It is broken down into seven
major sections:

1. Scope, Purpose and Application.
2. A General Description of Shears.
3. Definitions.
4. Construction, Reconstruction and Modification.
5. Safeguarding.
6. Care.
7. Use.

The effective dates of the various sections are very important to the shear user because they are virtually certain to be adopted by OSHA. For the majority of shear users, Sections 4, 5, 6, and 7 are the most important.

For new shears manufactured for installation in the United States, the rules spelled out in Sections 4 and 5, which cover construction, reconstruction, modification, and safeguarding became effective August 16, 1974, as far as ANSI is concerned and will probably have become effective as OSHA Standards by the time this revised edition of the Press Brake and Shear Handbook is published. They should not present any real problems. Most major shear builders have been designing their equipment to meet B11.4 for the last several years. Many shear builders, it should be noted, have served on the B11.4 committee that developed the standard.

Existing installations must be modified by the employer to the extent necessary to bring them into compliance with the standard by August 16, 1976. The August 16, 1976, deadline also applies to a very important subsection of Section 7 that will require the employer to furnish hand tools so that it will not be necessary for the operator to place his hands within the point of operation. (The point of operation is defined as the area in the shearing plane between the upper and lower blades and the area beneath the holddowns.) A similar no-hands-in-the-die-area provision in the OSHA Standard covering power presses caused enormous problems in the stamping industry. As applied to shears, the problem should be less severe.

National Standard B11.4 is written in such a manner that the responsibilities of the shear builder and the shear owner are clearly spelled out in Section 4. The section covers construction, reconstruction, and modification, as noted, and it will not create a heavy burden for either builder or user. The controls on many older shears will have to be modified and doubtless a number of control "packages" will be placed on the market as has already happened in the case of the press industry. An excellent feature of the standard is that it recognizes that there are substantial differences between an automated, long-run operation and a manually fed, short-run operation. Users will have to make changes in some areas such as foot pedals and two-hand controls, and in the installation of devices to prevent unintentional operation and repeat strokes. No great cost should be involved in meeting the provisions of Section 4 and the alloted time for compliance is adequate.

Section 5, which covers safeguarding, will present some problems. It requires that shears be equipped with either a point of operation guard, a point of operation device, or a point of operation awareness barrier. The builder is responsible, in the case of new equipment, and the user is responsible in the case of existing installations. This section may cause some problems and involve some expense. It's also a bit difficult to understand for semantic reasons. The standard recommends fixed point of operation guards as the most effective form of operator protection. The standard recognizes, however, that this type of safeguarding is not always feasible and recognizes point of operation devices as an alternative.

There are several types of point of operation devices. They include pullbacks, light curtains, RF barriers, and so on. Two-hand controls, incidentally, are not acceptable as point of operation devices, which is unfortunate because, properly used, they are very effective. Exactly what the standard means by a point of operation awareness barrier—this is a suggested alternative safeguarding method—is not quite clear. It apparently refers to a barrier that can be moved to permit entry of the metal to be sheared, but which requires a conscious effort on the part of the operator. The standard spells out the exact clearances permitted with both fixed guards and point of operation awareness barriers.

Section 5 contains a subsection that covers safeguarding of the rear of the shear. The section requires the employer to set up safe working practices and to provide safeguards to protect the shear helper or other persons behind the shear. The standard suggests a number of approaches to this problem, but does not set down any hard and fast rules.

Sections 6 and 7 cover care and use respectively. The former requires the shear builder to furnish adequate installation and operating instructions. It requires the shear owner to conduct training programs for maintenance personnel, to set up an inspection and maintenance program and to keep records thereof. Section 7 is rather general in nature. One of its provisions, and a good one, requires that employer instruct the operator in safe working methods and that he adequately supervise operating personnel to insure that they follow the rules. There is a very good outline of shear operator responsibilities and procedures. It is not a part of the basic standard but is included as an explanatory note.

An Analysis of B11.3

National Standard B11.3 covers power press brakes—mechanical and hydraulic. It does not include hand brakes, tangent benders, folding machines and pan or apron brakes. It also specifically excludes straightside presses with four or more gib points of the type often used with unitized tooling. These machines are covered by the mechanical power press standard B11.1-1971.

The standard recognizes two classes of power press brakes—general-purpose machines and special-purpose press brakes—and covers both mechanical and hydraulic types. Simply expressed, general-purpose machines are those operated by only one operator, working from the front of the machine, with a single operating control. Special-purpose machines are those that can be operated by one or more operators from the front of the machine, each having his own control station. The special-purpose category also includes press brakes with bolster extensions, press brakes with special machined beds used for multiple die stations, and a number of additional machines.

Exactly why the distinction is made in the standard is not precisely clear. For example, special-purpose dies may not be installed in a general-purpose press brake even if bolster extensions are added. And a two-speed drive that provides for fast approach, slow form, and fast return stroke automatically rates the press brake as a special-purpose machine even though such drives are so widely used that they are offered as a standard feature by several press-brake builders.

The foreword to B11.3 contains the comment that the job of developing a standard for press brakes is complicated by the wide variety of machines available and the many manufacturing methods with which they are used. This is certainly true and it reflects credit on the B11.3 subcommittee for having developed a practical standard in spite of the complexity of the task.

National Standard B11.3-1973 is divided into six major sections as compared with the seven sections in the B11.4-1973 standard. The sections are:

1. Scope, Purpose and Application.
2. Power Press-Brake Descriptions.
3. Definitions.
4. Construction, Reconstruction, and Modification.
5. Care.
6. Use.

It will be noted that a section on safeguarding, found in almost all standards in the B11 series, has been omitted. The reason, of course, is that safeguarding equipment such as barrier guards, most point of operation devices, and so on, are not compatible with the operation of a press brake. Since this is the case, the thrust has been in the direction of better, more reliable controls, improved operator training, and so forth.

As is the case with the safety standard for shears, the press brake standard does not impose impossible burdens on the builders or the owners of press brakes. Like the shear standard it provides the very important service of defining responsibilities. The manufacturer is responsible for building new equipment in accordance with the standard. The responsibility for modification or reconstruction rests with the person doing the job. The owner (employer) is responsible for eliminating hazards at the point of operation.

Most of the requirements spelled out in the standard are reasonable, represent good safety practice, and will not be difficult to comply with. For example, compression springs used to actuate a friction-type brake on a mechanical press brake must be either guided on a rod or in a hole to eliminate a hazard should they break under load. The brake capacity must be sufficient to stop the ram when the operator lifts his foot from the pedal and to hold the ram at any point in the stroke. It would be difficult to quarrel with that requirement. Nor can anyone quarrel with the requirement that foot controls be protected so that they cannot be actuated by falling objects.

Also a common-sense requirement is for a main disconnect switch, capable of being locked only in the "off" position for each press brake. The purpose is to completely isolate the press brake from the main power supply for maintenance, service, and so on. Yet another subsection of the standard requires that all mechanical press brakes be equipped with a hand rail mounted on the frame or ram. The purpose is to help the operator to maintain his balance and also to occupy his hands when working the equipment with a foot pedal.

Some of the requirements for controls seem somewhat restrictive and in some cases redundant, but on the whole B11.3 will not be hard to live with. There is one subsection under "Use" that merits immediate attention. Under the general heading "Die Design, Construction, Procurement and Modification" it requires that the employer, prior to February 15, 1976, institute die design policies and procedures that will make it unnecessary, after that date, for the employee to place his hands or fingers within the point of operation. The standard suggests a number of approaches that can be taken to this requirement.

Copies of ANSI B11.3-1973 and B11.4-1973 are available at $2.50 each from the American National Standards Institute, 1430 Broadway, New York, New York 10018. Because they are almost certain to be adopted verbatim as OSHA standards, they should be obtained and carefully studied by all users of shears and press brakes.

The Importance of Safety Training

Almost all of the OSHA published standards emphasize the importance of operator training, regardless of the type of machine. Innumerable citations have been issued by OSHA compliance officers based on the lack of formal training programs and printed training material. Machine tool builders are now required to furnish comprehensive instruction manuals dealing with the installation and use of their equipment. It is important to back this material up with less complicated publications comprehensible to employees of every degree of education. Cincinnati, Incorporated, has published brief manuals covering mechanical and hydraulic shears and mechanical and hydraulic press brakes. Portions of the manual covering mechanical press brakes are reproduced in the following pages. Similar manuals covering many types of equipment can be prepared by any competent manufacturing engineering department, adapting appropriate sections to cover local and internal conditions. Preparation of such manuals and issue (with receipts and certification that the employee has read them) is a strongly recommended procedure. Such a procedure is generally considered by OSHA compliance officials as indicating genuine management interest and effort toward compliance with federal safety standards. The manual published here can be reproduced without further permission but it would best be used as a guide in preparing internal manuals, providing a similar depth of detail.

MECHANICAL PRESS BRAKE SAFETY MANUAL

Never operate a press brake or perform maintenance on it without proper instructions and without first reading and understanding the official Operating and Maintenance Manual furnished with the machine. This informal manual is not intended as a substitute for the official manual but rather to supplement it and to offer additional general safety suggestions.

Keep Clear of the Work Area

The purpose of a press brake is to bend metal and it is obvious that this same capacity will sever arms, fingers, or any part of the body that is in the ram area when the ram is actuated.

During operation, all parts of your body must be completely clear of the work area. If operation by more than one man is required and a foot switch or two palm-operated push buttons are not provided for each operator, only one man should have the responsibility for actuating the machine. It should be his responsibility to see that not only his own body is clear of the work area and all moving parts, but that his coworkers are also clear and entirely visible in a safe area before actuating the press brake.

During setup, maintenance or other work on the machine that requires manipulation within the work area, the ram should be at the bottom of the stroke or should be blocked up so that the dies cannot close, the flywheel should be stopped, and the power supply completely disconnected.

Concentrate on Your Job

Daydreaming, worrying about other problems, or other improper operation of any machine could cripple you for life. Operating a press brake requires your complete attention. Talking, joking, or participating in or watching horseplay could result in physical injury to you—which is certainly nothing to joke about. Watch what you are doing and concentrate on your job.

The Importance of Neatness

Keep the floor of your work area clear of scrap and trash that could cause you to trip or stumble. Put scrap in the proper containers and keep stock and finished work neatly arranged. Be sure slippery surfaces are properly cleaned up. Stumbling and slipping have resulted in painful and sometimes fatal injuries.

Put all tools away when you are not using them. Only the part you are working with should be on the machine when it is operating. Even a screwdriver can become a deadly projectile if left on the press brake bed or lower die.

Use the Proper Tools

Use the proper tool when working on the press brake. An improper tool might slip and cause severe lacerations. When making repairs on the machine disconnect the power source and be sure that the ram is at the bottom of the stroke or is blocked in place with the flywheel stopped.

Loose or flowing clothes may be comfortable but they should not be worn around machinery. Keep jewelry to a minimum. That link ID bracelet you got for Christmas could cost you your hand or finger.

Make a Pre-Start Check

Before operating the press brake look to see that it is in the proper condition. Are the dies worn? Is the floor area clear and clean? Are your tools put away where they belong? Is the stock neatly arranged? Are all covers and guards in place? Is the machine securely anchored to the floor? Are all nuts, bolts and screws tight? Is everything in proper operating condition? This check takes only a few moments but it is very important. Commercial airlines have the best maintenance services possible but the flight engineer or the pilot makes a similar check of his aircraft before every takeoff. If you see anything that strikes you as unsafe, report it to your supervisor at once and do not start operations until the situation is corrected.

Watch Out for Overloading

Your press brake has a maximum capacity as indicated on the capacity plate attached to the machine. Normally, work scheduled for a press brake will be well within the capacity of the equipment. Sometimes a slipup may occur. If you feel that this has happened, don't try to run the job. Ask the supervisor or foreman to check the tonnage charts to be sure that the job belongs on your particular machine. By overloading, you can damage the tooling, the drive train, or the press brake bed and ram. When using short or small area dies, the force must be reduced to prevent damage. Too much pressure can also cause a die to rupture and cause injury.

Rules for Safe Operation

1. Know how to safely operate and adjust your press brake. Review the Operating and Maintenance Manual provided by the builder.

2. Maintain proper lighting levels and eliminate light glare to prevent eye strain and eye fatigue.

3. Protect your eyes from flying pieces of metal by always wearing your safety glasses.

4. Always wear safety shoes. A heavy or pointed piece of stock could fall and cause serious and painful injury to your foot. (This is one of the most common industrial accidents.)

5. Wear snug-fitting hand and arm protection when handling rough or sharp-edged stock.

6. Keep the die area free of loose tools and materials. When placing stock in the machine for forming, be certain that the gages and stops are correctly set and that the edge of the stock is set flush with the gages.

7. Stand clear of the workpiece with thumbs and fingers beneath the workpiece and arms slightly extended to avoid being hit in the face if the stock whips upward as the bend is made. With some dies, the whip is down. Be sure that you know how the workpiece will react to the bend being made.

8. Releasing the treadle or electrical foot control of your Cincinnati mechanical press brake will stop downward travel of the ram in case of emergency. (Note: This is an ANSI B11.3-1973 requirement on all mechanical press brakes. The B11.3-1973 standard also requires provision of an emergency stop button.)

9. When you leave the machine, place the ram at the bottom of the stroke or place safety blocks in position under the ram, engage the treadle lock or turn the foot control switch OFF, and shut the power OFF even if you will be away for only a few moments.

10. Have the levelness of your machine checked at least once a month. The proper procedure is described in the official Operation and Maintenance Manual provided by the builder.

11. Check the alignment of the dies before operating the machine after the dies have been changed or if the machine has been idle overnight.

Improper alignment can cause chipping and flying chips can cause lacerations and eye injuries.

12. Report any cuts, bruises, and all other injuries to your supervisor or the medical department immediately. They are the best judges of how serious or minor your injury may be.

Changing Punches and Dies

Removing, transferring and setting up dies is hazardous and should be done under proper supervision by experienced set-up men. Improper handling techniques can cause muscle strains and more serious injuries.

Whenever punches or dies are removed from or installed in the machine, always set the ram at the bottom of the stroke and lock the treadle. From this point, the ram can only move UP if the treadle is accidentally depressed. Properly designed mechanical press brakes have register marks on the eccentric shaft to indicate when the ram is at the bottom of the stroke.

1. For the safe removal of punches and dies, use the following procedure:
   a. Clear the work area of all stock, containers, tools, and other equipment.
   b. Place safety block on top of lower die.
   c. Clean both upper and lower dies using a bench brush and finally wipe clean with a cloth. Remove safety block after all work in the die area is finished and jog the machine to the bottom of the stroke.
   d. Jog the ram adjusting motor until the upper and lower dies are completely closed. Loosen the ram clamp bolts to release the upper die.
   e. Jog the ram adjusting motor up to release any pressure between the dies. About ⅛ inch of motion is sufficient.
   f. Slide the upper die slightly more than half way out of the machine.
   g. Attach a lifting sling at the middle of the die and lift the upper die from the lower die.
   h. Now remove the upper die from the machine with a lift and transfer it to the tool storage area, making sure that the working surfaces of the die are protected from nicks and scratches. Be sure that dies that are top-heavy are blocked to prevent tipping.
   i. Loosen the bolts holding the lower die to the filler block.
   j. Slide the lower die out of the machine. Lift in the same manner as used for the upper die and transfer to the tool storage area.
2. Transfer dies using the proper techniques for the weight of the die being handled.
   a. Very light dies (up to 50 lbs) can be carried manually or transferred to a die truck.

  b. Dies weighing over 50 lbs should be handled with a hoist.

  c. If the punch or die has tapped holes for lifting attachments, be sure that the proper size bolts are used. A bolt smaller in diameter than the tapped hole will slip out and can cause serious injury. If no lifting attachments are provided, use only approved rope slings so that the die will not be nicked or scratched.

  d. Lift the dies high enough to clear any obstructions but no higher.

  e. Stay clear of dies and punches while they are being transferred, particularly when dies are being lifted. If a die should slip, serious injury, including loss of a hand, a foot, or even your life could result.

3. When setting up punches and dies:

  a. Make sure that the top of the press brake bed and the top of the filler block are clean. These should be cleaned and wiped off only when the ram is at the bottom of the stroke.

  b. Inspect the die for chips, cracks, and so on. Wipe them off with a clean cloth.

  c. With the ram at the bottom of the stroke and the filler block in place but not bolted tight, place the lower die on the filler block in such a way that the load will be centered between the side frames if possible. Tighten the die clamping set screws in the filler block. The die should sit firmly on its supporting shoulders. Locate the filler block so that the die vee is centered above the slot in the bed and temporarily tighten the clamp bolts.

  d. Adjust the ram by means of the ram adjusting motor so that just enough space remains for the upper die. Slide the upper die into place and tighten the die clamp bolts securely so that the upper die will not drop out. Adjust the ram upward to free it from the lower die and place safety blocks on the lower die. Recenter the lower die by sight. Now remove the safety blocks and adjust the ram downward to seat the die in the die clamps and complete the final tightening of the upper die clamp bolts.

  e. Adjust the ram upward to metal thickness clearance at the slopes of the die. Adjust the alignment of the filler block until the clearances are alike when the filler block is bolted securely to the bed. Check at both ends of the die with feeler gages or strips of metal of proper thickness. This completes the alignment setup.

  f. The ram must now be adjusted to produce an acceptable part. This may require a different setting at one end from the other to compensate for errors in dies and differences that may occur in wear on dies. Some jobs will require shimming of dies to correct for machine deflections. If the load is too light to deflect the bed and ram to a parallel condition, shims may be required.

g. A definite change in the angle of the bend will be noted if the ram adjustment or shimming is changed only a few thousandths of an inch. In like manner, shims may be only a few thousandths of an inch in thickness. The "tapering" procedure is important in gaining satisfactory results.

h. Make test bends using the same type and thickness of material that will actually be used on the job. Adjust the ram setting until a satisfactory bend is made.

i. When a satisfactory setup has been made, all information such as dies used, filler block used, ram adjustment reading, ram tilt readings, gage setting dimensions, shims (if used) and other notes that would be useful in repeating the job with a minimum of setup time should be recorded.

j. Now that the job is ready to run, make a complete Pre-Start Check before starting into production.

INDEX

Press Brakes

INDEX (con'td)

Shears

ABOUT THE AUTHOR

Harold R. Daniels is currently a consulting editor with *Metal Stamping* magazine. Over the past fifteen years, he has been an active member on the Technical Research and Standards Committee of the American Metal Stamping Association. He previously served as Senior Associate Editor of *Metalworking* magazine and is a recipient of the Presteel Award for his outstanding contributions to the industry. Mr. Daniels has published numerous articles in trade journals, and is the author of *Mechanical Press Handbook*.